Principles and Practices of
Bioslurping

Matthew C. Place
Christopher T. Coonfare
Abraham S.C. Chen
Ronald E. Hoeppel
Stephen H. Rosansky

BATTELLE PRESS
Columbus • Richland

Library of Congress Cataloging-in-Publication Data

Principles and practices of bioslurping / Matthew C. Place ... [et al.].
p. cm.
Includes bibliographical references and index.
ISBN 1-57477-107-8 (case bound : alk. paper)
1. Oil pollution of soils. 2. Soil remediation--Technological innovations. 3. Groundwater--Purification--Technological innovations. 4. Soil bioventing. I. Place, Matthew C., 1968-

TD879.P4 P755 2001
628.5′2--dc21

2001025995

Printed in the United States of America

Battelle Press
505 King Avenue
Columbus, Ohio 43201-2583, USA
1-800-451-3543 or 614-424-6393
Fax: 614-424-3819
Homepage: http://www.battelle.org/bookstore
E-mail: press@battelle.org

Foreword

Leaks and spills from pipelines and underground and aboveground storage tanks at fueling terminals and service stations may result in extensive contamination to the aquifers and vadose (or unsaturated) zones beneath them. Leaked and spilled petroleum hydrocarbons (such as jet fuels, gasoline, diesel, heating oils, and crude oils) may pose direct and indirect safety hazards and cause adverse effects to human health and the environment. Therefore, it is desirable to remove free-phase and dissolved plumes and remediate contaminated vadose zones, especially in cases where contaminant plumes are situated underneath densely populated areas.

Over the past 10 years, bioslurping has evolved from a primarily experimental technology to a preferred method for remediating sites contaminated with free-phase product. By combining vacuum-assisted free-product recovery with bioventing and soil vapor extraction, bioslurper systems simultaneously recover free product and remediate the vadose, capillary, and saturated zones. Specifically, bioslurper systems use an aboveground vacuum pump to create enough vacuum and airlift to withdraw light, nonaqueous-phase liquids (LNAPLs) from the subsurface, along with soil gas and relatively small amounts of groundwater. The accompanying pressure gradient promotes the migration of the free product to the well via more permeable flow pathways, and avoids the trapping of free product below the water table by preventing rebound of the water table. Thus, bioslurping offers improvement on more traditional free-product recovery technologies, which tend to rely on hydraulic gradients and cones of depression to promote free-product movement.

This manual provides the reader with a detailed description of bioslurping theory, practice, and system design. It begins with a description of the principles of free-product migration through the subsurface and methods for enhancing the recovery of free product. Chapter 2 describes methods for characterizing the site and evaluating free-product recovery potential at the site. In addition, methods for determining the potential for groundwater contamination from free-product sources are presented, and a protocol for free-product recovery testing is provided so readers of this book can determine if bioslurping is an appropriate free-product recovery technology for use at their sites. Chapter 3 describes the design, installation, and monitoring of pilot-scale systems; the chapter also presents methods for determining

LNAPL recovery potential, vadose zone and LNAPL radii of influence, and the bioventing potential of the bioslurper at the site. Chapter 4 provides recommendations for full-scale design of a bioslurper system using the data generated from a pilot test area, and includes a description of the operation and maintenance of full-scale bioslurper systems (including treatment of aqueous and vapor discharge streams). Finally, Chapter 5 presents an exit strategy for LNAPL recovery and site closure. This example exit strategy has been designed to accommodate the various site-specific conditions that readers of this book are likely to encounter at their sites; also, the strategy can be modified in order to meet site- and/or project-specific regulatory goals.

The authors would like to thank the following individuals for their contributions at various bioslurper demonstration sites and for advancing the technology: Dr. Larry Smith, Charles Cummings, Dr. Andrea Leeson, Dr. Godage Wickramanayake, Craig Payne (especially for his contributions to the pre-pump separation systems), Eric Drescher, Jim Gibbs, and Dr. Rob Hinchee (Battelle); Jeff Kittel (for doing much of the initial work of developing the technology) (Midwest Research); Art Fisher and Dr. Raj Krishnamoorthy (Naval Air Station Fallon); Patrick Haas and Jerry Hansen (Air Force Center for Environmental Excellence); Erlingur Jonasson (Naval Air Station Keflavik); Dr. Ralph Baker (ENSR); Mike Hawthorne (H2A Ltd.); John Talley (Naval Facilities Engineering Service Center); and Sten Pearson (Brewer Environmental).

We also appreciate the dedication of the following individuals who operated bioslurper systems, collected data, and developed the bioslurping technology: Roger Johnson, Greg Headington, Mike Woolfe, Jon Eastep, Shane Walton, Sarah McCall, Travis Williamson, Kip Hewahewa, Julie Kramer, Emmett Black, Lydia Cumming, and Sam Serpa (Battelle). These people endured Alaska in the winter and Georgia in the summer to conduct the tests and demonstrations of bioslurper systems, and their efforts truly are appreciated. Also, thanks to Jerry Tompkins for purchasing and tracking the materials used at many bioslurper test sites.

Finally, the authors would like to thank Mark Hendershot for his editorial review and Loretta Bahn for her desktop publishing of this book, Pat Greene for text processing the initial versions of this bioslurping guidance manual, Vic Saylor and Dean Sutton for preparing many of the figures presented here, David Reed for designing the cover art, and Joe Sheldrick for assisting in the preparation of this document.

Contents

Foreword ... iii
Contents ... v
Figures ... viii
Tables ... ix
Appendices ... ix
Acronyms and Abbreviations ... xi

1 Introduction ... 1
1.1 LNAPL Recovery Technologies ... 1
1.1.1 Skimming ... 2
1.1.2 Drawdown Pumping (Single- and Dual-Pump) ... 5
1.2 Bioslurping ... 6
1.2.1 Bioslurper Technology Description ... 6
1.2.2 Principles and Theory of LNAPL Migration and Vacuum-Enhanced Recovery ... 7
1.2.2.1 LNAPL Migration ... 7
1.2.2.2 Vacuum-Enhanced Recovery ... 14
1.2.3 Bioventing Component of Bioslurping ... 18
1.2.4 Soil Vapor Extraction Component of Bioslurping ... 19
1.3 References ... 20

2 Site Characterization and Evaluation ... 23
2.1 Site Characterization ... 23
2.1.1 LNAPL Release History ... 23
2.1.2 Areas/Extent of Characteristics of Contaminants ... 24
2.1.2.1 Interface Measurements ... 24
2.1.2.2 Baildown Testing ... 25
2.1.2.3 Soil-Gas Survey ... 27
2.1.2.4 In Situ Respiration Testing ... 29
2.1.2.5 LNAPL Plume Delineation and Volume Estimation ... 29
2.2 Characterization of Site Geology and Hydrogeology ... 32
2.2.1 Site Geology Analyses ... 32
2.2.2 Site Hydrogeology ... 33
2.2.3 Fluid Characterization ... 36

2.3 Evaluation of Site Characterization Data 38
2.4 References 39

3 Pilot Testing 41
3.1 Permitting for Bioslurper Pilot Testing 42
3.1.1 Vapor Discharge Permits 42
3.1.2 Aqueous Discharge Permits 42
3.1.3 Well Installation Permits 43
3.1.4 Other Permits 43
3.2 Bioslurper Extraction Wells 43
3.3 Soil-Gas Monitoring Points 45
3.3.1 Locations of Monitoring Points 46
3.3.2 Monitoring Point Construction 48
3.4 Baseline Measurements 50
3.5 Mobilization, Installation, and System Shakedown 50
3.5.1 Pilot-Scale Bioslurper System 51
3.5.2 Site Mobilization 52
3.5.3 System Installation and Shakedown 53
3.6 LNAPL Recovery Testing 53
3.6.1 Simulated Skimmer Testing 55
3.6.2 Bioslurper Extraction Test 55
3.6.2.1 LNAPL Recovery Rate and Volume 56
3.6.2.2 Groundwater Recovery Volume 57
3.6.2.3 Vadose Zone Radius of Influence 57
3.6.2.4 Stack-Gas Volume 58
3.6.2.5 Sampling and Analyses for Process Monitoring 59
3.6.2.6 Soil-Gas Monitoring 62
3.6.2.7 LNAPL Thickness and Groundwater-Level Monitoring 62
3.6.3 Pump Drawdown 64
3.6.4 Simulated Soil Vapor Extraction 65
3.7 Data Reduction and Results Interpretation 66
3.7.1 LNAPL Thickness Monitoring 66
3.7.2 LNAPL Radius of Influence 67
3.7.3 LNAPL Extraction Tests 69
3.7.4 Soil-Gas Monitoring and Shutdown Respiration Testing 72
3.8 References 72

4 Full-Scale Bioslurper Design and Construction 75
4.1 Extraction Well Spacing and Design 75
4.2 Drop Tube and Extraction Manifold Sizing 77
4.3 Liquid Ring Pump 78
4.4 Oil/Water Separation 82
4.4.1 Separation Using a Knockout Tank 82
4.4.2 In-Well Oil/Water Separation 84
4.4.3 Separation Through Physical Removal 85
4.4.4 Separation Using Coalescers 85
4.5 Groundwater Effluent Treatment 86
4.6 Stack-Gas Treatment 90
4.7 Safety Switches and Equipment 90
4.8 Life-Cycle and Cost Analysis 95
4.9 Design Considerations for System Operation 97
4.9.1 Noise Abatement 98
4.9.2 Power Availability 99
4.9.3 Pedestrian and Vehicular Traffic 99
4.9.4 Fluctuating Groundwater Tables 100
4.9.5 Cold Weather Operation 101
4.10 References 102

5 Bioslurper System Operations and Maintenance 103
5.1 Troubleshooting and Care of the Liquid Ring Pump 103
5.2 Operation of Oil/Water Separator 108
5.3 Troubleshooting and Care of Water Treatment Systems 109
5.3.1 Chemical Treatment and Dissolved Air Flotation 109
5.3.2 Air Stripping 110
5.3.3 Granular Activated Carbon/Clay Anthracite Adsorption Media 111
5.3.4 Gravity Settling Tanks 112
5.3.5 Reinjection 113
5.4 Troubleshooting and Care of Stack-Gas Treatment Systems 113
5.4.1 Reinjection/In Situ Biodegradation 113
5.4.2 Granular Activated Carbon 114
5.4.3 Catalytic/Thermal Oxidizer 114
5.4.4 Internal Combustion Engine 115
5.5 Health and Safety Issues 116

6 Site Closure 119
6.1 References 121

Figures

1-1. Comparison of pumping, skimming, and bioslurping methods for LNAPL recovery 4
1-2. Progression of an example LNAPL release 8
1-3. Distribution of air, hydrocarbon, and water in a porous medium and an observation well 9
1-4. Hypothetical relative permeability curves for water and LNAPL in a porous medium 14
2-1. Baildown test record sheet 26
2-2. Typical setup for monitoring soil gas 28
3-1. Diagram of a typical bioslurper well 46
3-2. Conceptual configuration of soil-gas monitoring points and bioslurper well 47
3-3. Diagram of a typical soil-gas monitoring point 49
3-4. Trailer-mounted, pilot-scale bioslurper unit 51
3-5. Bioslurper pilot test shakedown checklist 54
3-6. Configuration of drop tube and wellhead during simulated skimmer test 56
3-7. Configuration of drop tube and wellhead during bioslurper extraction test 57
3-8. Typical flow calibration curve for the bioslurper vapor discharge pitot tube, 3-inch, schedule 80 PVC stack 58
3-9. Diagram of the in situ interface probe setup 63
3-10. Configuration of drop tube and wellhead during simulated drawdown recovery test 64
3-11. Decision tree for selecting an LNAPL-recovery technology 67
3-12. Simulated flow vectors for (A) the oil phase and (B) the air phase 70
4-1. Stepwise approach to full-scale bioslurper design 76
4-2. Recommended extraction well layout 77
4-3. Flow patterns at superficial liquid (V_{SL}) and gas (V_{SG}) flow velocities 79
4-4. Effect of multiple wells on soil-gas extraction rate 81
4-5. Vacuum-resistant separator 83
4-6. Dual-drop tube design for in-well separation 84
4-7. Oil/water separator 87
4-8. Concentration ranges for vapor treatment methods 91
4-9. Electrical configuration of an example annunciator panel 95
4-10. Typical mass removal curve (rate vs. time) 96
6-1. Generic decision diagram for bioslurper exit strategy and site closure 120

Tables

2-1. Parameters to be measured for the in situ respiration test 30
2-2. Ranges of porosity for various soil and rock types 36
2-3. Characteristics of unweathered fuels 37
3-1. Schedule of activities...55
3-2. Suggested sampling and analytical methods 60
4-1. Aqueous treatment and disposal options 88
4-2. Water treatment technologies 89
4-3. Stack-gas treatment technologies 92
5-1. Troubleshooting guide for the liquid ring pump unit 104
5-2. Exposure limits for chemical substances.......................................117
5-3. Potential hazards and recommended preventative measures..........118

Appendices

A Bioslurping Implementation Cost-Estimating Guide 123
A.1 Introduction and Scope 123
A.2 Bioslurping Installation, Operations, and Maintenance 123
A.2.1 Construction of Mobile Pilot Unit 123
A.2.2 Pilot-Scale Installation and Testing................................ 127
A.2.3 Full-Scale Installation and Testing.................................. 132
A.2.4 Operations and Maintenance 136
A.3 Other Costs 140
A.3.1 Design Costs... 140
A.3.2 Documentation Costs.. 140
A.3.3 Site Closure (Sampling and Analysis) Costs................... 142
A.3.4 Contingency Costs.. 143
A.3.5 General and Administration, Overhead, and Fee 143
A.4 Conclusions 143

B Water Treatment Technologies 145
B.1 Chemical Treatment and Dissolved Air Flotation...................... 145
B.2 Air Stripping.. 150
B.3 Granular Activated Carbon.. 151
B.4 Clay/Anthracite Adsorption Media 152
B.5 Gravity Settling Tanks... 153
B.6 Reinjection... 154
B.7 References 154

C Stack-Gas Treatment Technologies 155
C.1 GAC Adsorption 155
C.2 Biofiltration 156
C.3 Reinjection/In Situ Biodegradation of Vapor Emissions 156
C.4 Destruction in a Catalytic Oxidizer 157
C.5 Thermal Oxidation Flame Incineration 159
C.6 Flameless Incineration 159
C.7 Destruction in an Internal Combustion Engine 160
C.8 References 164

Index 167

Acronyms and Abbreviations

ACGIH	American Congress of Governmental and Industrial Hygienists
AFB	Air Force Base
AFCEE	Air Force Center for Environmental Excellence
ASTM	American Society for Testing and Materials
bgs	below ground surface
BTEX	benzene, toluene, ethylbenzene, and xylenes
cfh	cubic feet per hour
cfm	cubic feet per minute
CFR	Code of Federal Regulations
CQC	Contractor Quality Control
CRF	chemical reaction/flocculation
CSS	Coastal Systems Station
DAEM	Draper Aden Environmental Modeling, Inc.
DAF	dissolved air flotation
GAC	granular activated carbon
GC/FID	gas chromatography/flame ionization detection
gpm	gallon(s) per minute
hp	horsepower
ICE	internal combustion engine
ID	inside diameter
JP	jet propellant
LEL	lower explosive limit
LIF	laser-induced fluorescence
LNAPL	light, nonaqueous-phase liquid
LRP	liquid ring pump

MDL	method detection limit
MP	monitoring point
MTBE	methyl-*tert*-butyl ether
NA	not applicable
N/A	not available
NAPL	nonaqueous-phase liquid
NEMA	National Electrical Manufacturers Association
NFESC	Naval Facilities Engineering Service Center
NPDES	National Pollutant Discharge Elimination System
O&G	oil and grease
O&M	operations and maintenance
OD	outside diameter
OSHA	Occupational Safety and Health Administration
OVA	organic vapor analyzer
OWS	oil/water separator
PAH	polycyclic aromatic hydrocarbon
PEL	Permissible Exposure Limit
ppbv	parts per billion by volume
PPE	personal protective equipment
ppm	parts per million
ppmv	parts per million by volume
psi	pound(s) per square inch
psia	pound(s) per square inch absolute
psig	pound(s) per square inch gauge
PVC	polyvinyl chloride
QAPP	Quality Assurance Program Plan
RCRA	Resource Conservation and Recovery Act
ROI	radius of influence
RPM	Remedial Project Manager
rpm	revolutions per minute
scfm	standard cubic feet per minute
SFRWB	San Francisco Regional Water Board
SVE	soil vapor extraction
TCLP	Toxicity Characteristic Leaching Procedure
TLV	threshold limit value

TPH	total petroleum hydrocarbons
TSS	total suspended solids
U.S. EPA	United States Environmental Protection Agency
UST	underground storage tank
V_{SG}	superficial gas velocity
V_{SL}	superficial liquid velocity
VOA	volatile organic analysis
VOC	volatile organic compound
WWTP	wastewater treatment plant

Principles and Practices of

Bioslurping

1

Introduction

One of the primary steps in remediation of petroleum hydrocarbon-contaminated sites involves source removal. For example, the U.S. Environmental Protection Agency's (U.S. EPA's) Directive on Monitored Natural Attenuation (U.S. EPA, 1997) requires implementation of source removal prior to remediation of the site by natural attenuation. Source removal consists primarily of removal of light, nonaqueous-phase liquid (LNAPL) from the subsurface. Effective implementation of any remedial technology at petroleum hydrocarbon-contaminated sites can be impaired significantly if the free product is not first removed. For example, 1 gallon of residual gasoline containing approximately 3% benzene potentially can contaminate millions of gallons of groundwater, not accounting for volatilization, adsorption, and biodegradation. Therefore, the primary goal of the source removal should be the expeditious elimination of free product to the maximum extent practicable. Past experience at sites contaminated with LNAPL indicates that the vacuum-enhanced bioslurper technology generally is more effective at achieving maximum free-product removal compared to conventional LNAPL removal technologies such as skimming and drawdown pumping (Leeson et al., 1995; Parker, 1996; Reisinger et al., 1993). The purpose of this manual is to present an overview of bioslurping compared to other conventional technologies and to document procedures for design and implementation of bioslurping at sites contaminated with free product.

1.1 LNAPL Recovery Technologies

Conventional LNAPL extraction technologies recover organic contaminants by physical collection of free-phase liquids. LNAPL floating on the water table in a well is withdrawn by the suction action of a pump or by selective collection using various skimming technologies. Successful collection requires a sufficient thickness of LNAPL in the well, which is replenished through the gravity-driven advective flow of LNAPL into the well. Different collection techniques or pump configurations may be used to optimize the LNAPL extraction under different in situ conditions. Three types of conventional extraction approaches are used: (1) skimming, (2) single-pump drawdown, and (3) dual-pump drawdown.

The LNAPL extraction device may be located in a well or a trench, depending on the depth to the water table, the extent of the contamination, and the topography of the site. Interceptor trenches and drains can be used at LNAPL-contaminated sites where the water table is near the surface. The trench is installed downgradient from the LNAPL source to intercept migrating free-phase liquids. LNAPL may migrate to the trench due to natural groundwater movement (skimming), or the flow of LNAPL and groundwater can be enhanced by pumping (single- or dual-pump drawdown) to lower the water table near the trench.

One of the most important characteristics of the trench is that it intersects the full geologic cross section. The ability to obtain flow across the full cross section is advantageous, especially at sites with discontinuous interbedded sands and clays. Nevertheless, trenches often are impractical at most sites. Because each trench must be excavated to a level below the lowest seasonal water table, a greater water table depth would significantly increase the challenge to maintain the stability of the trench walls. Further, the excavation brings contaminated soils to the surface that may require treatment and/or disposal. The excavation also may disrupt site access roads, utilities, or activities at the site.

Recovery wells are adaptable to a wider range of site conditions. A well recovery system may consist of a vertically installed well, or an array of wells, screened at the level of the water table and emplaced in the LNAPL plume. Recovery wells are placed at discrete locations, and may be placed to avoid utilities and roads. Furthermore, recovery wells may be used for a wide range of remedial activities. If LNAPL recovery is complete at a site, the wells generally may be used for bioventing and possibly soil vapor extraction (SVE).

1.1.1 Skimming

Skimming recovery systems use selective collection devices to collect LNAPL floating on the water table. One of the most widely used skimmers is the floating filter scavenger system that uses a floating filter (an oleophilic/hydrophobic mesh with a high affinity for nonpolar hydrocarbons) to allow passage of LNAPL and to reject polar molecules such as water. A mesh cylinder is designed to float in the LNAPL layer in a recovery well. LNAPL floating on the water table passes through the mesh while water is prevented from entering. The LNAPL runs down into a collection pot and is discharged periodically by air pressure to a central holding tank on the surface. The pressurization cycle may be controlled manually, by a timer, or by high- and low-level switches.

Shallow wells with low recovery rates can use rope wick or belt skimmers (Baker, 1995). The rope wick or belt skimmer uses a continuous loop of rope or belt made of an oleophilic/hydrophobic material. The rope or belt is strung through the LNAPL layer and up through a pair of compression rollers. The rollers provide the motive force for the rope or belt while squeezing out any retained LNAPL into a small container. LNAPL collected in the container is pumped periodically to a central holding tank. Large trench recovery points can be fitted with drum or disk skimmers that are too large to fit into a recovery well.

Skimmer systems withdraw little or no water and produce little or no drawdown, and thus have a limited pressure head to move LNAPL toward the recovery point (Figure 1-1). The passive action results in a relatively small radius of influence (ROI) from the recovery point. Moreover, the rate of recovery is low because the skimmer systems rely on the passive movement of LNAPL into the product recovery wells or trenches.

Despite the low LNAPL recovery rate, skimmers are popular due to the following desirable features:

- ❑ All of the skimmer designs allow recovery of LNAPL while recovering little or no water, which can reduce costs and operation constraints. When using any of the LNAPL recovery technologies, water brought to the surface must be treated and/or disposed of properly, thus increasing the permitting complexity and/or cost in many jurisdictions. As a result, high disposal costs and/or capacity limitations for wastewater management would favor selection of skimmers. The low water content of the LNAPL and the very small quantity of water withdrawn also reduce the size and complexity of skimmer fluid-handling systems.

- ❑ Skimmers will recover LNAPL when the thickness of the floating layer is too thin to allow efficient recovery with a pump. A skimmer can recover LNAPL even when the floating layer is less than ¼-inch thick.

- ❑ Skimmers are sturdy and mechanically simple with few moving parts. Further, skimmers are simple to install and easy to operate, and are available from a variety of vendors as complete package systems.

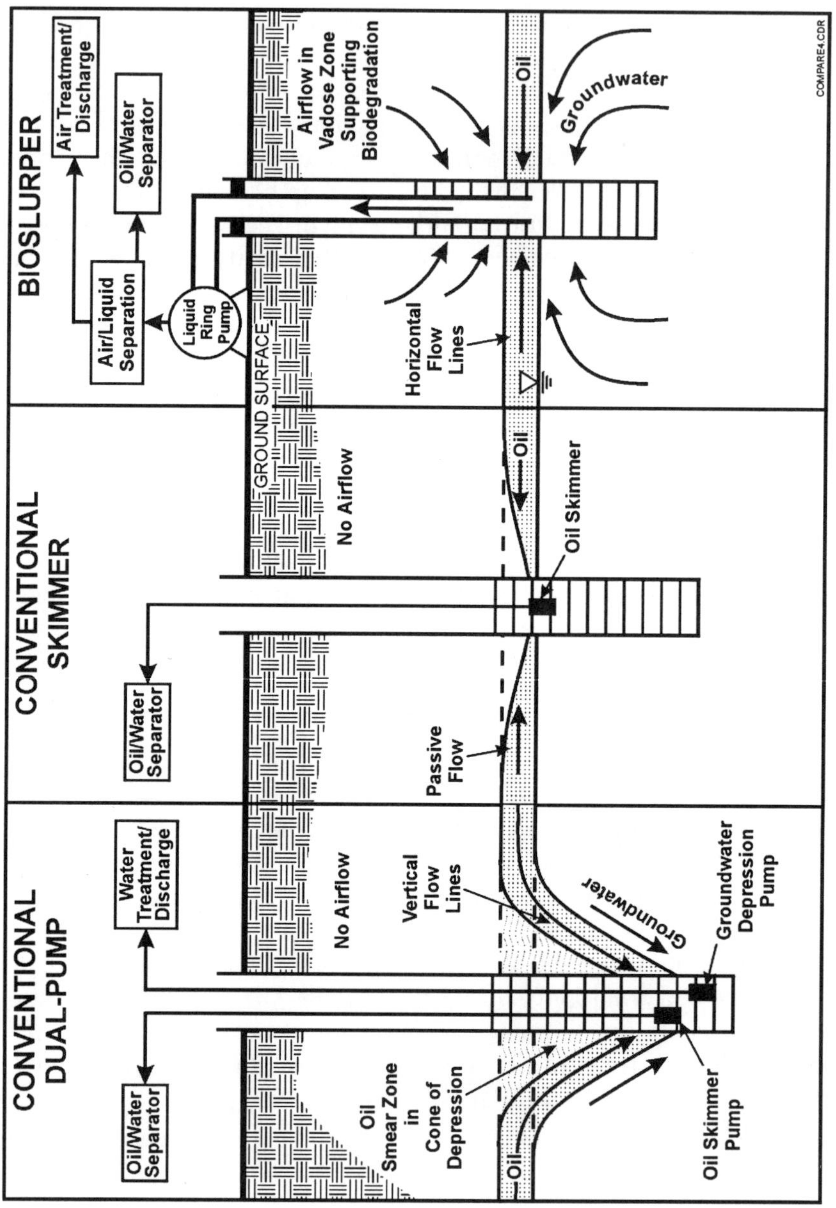

FIGURE 1-1. Comparison of pumping, skimming, and bioslurping methods for LNAPL recovery

1.1.2 Drawdown Pumping (Single- and Dual-Pump)

Pump drawdown LNAPL recovery systems are designed to pump LNAPL and groundwater from recovery wells or trenches. Pumping removes groundwater, establishing a cone of depression near the extraction area (Figure 1-1). The lower groundwater level in the vicinity of the well produces a gravity head that induces flow of LNAPL toward the well and increases the thickness of the LNAPL layer in the well. Each foot of groundwater depression provides a driving head equivalent to a pressure difference of 0.43 pound per square inch (psi). In most cases, the cone of depression will increase LNAPL recovery rates.

Pumping may be accomplished with one or two pumps. In the single-pump configuration, one pump withdraws both groundwater and LNAPL. The dual-pump configuration uses one pump located below the water table to remove groundwater and a second located in the LNAPL layer to recover LNAPL. A single-pump system reduces capital and operating costs and allows simpler control and operation. The system, however, produces a stream of water and LNAPL mixture that must then be separated. A dual-pump system often withdraws a large quantity of groundwater, which may need to be treated before disposal.

Drawdown pumping is effective for LNAPL recovery when the aquifer has moderate to high hydraulic conductivity and the LNAPL layer is thick. An aquifer with high hydraulic conductivity offers less resistance to the flow of LNAPL into the well. A thick layer of LNAPL allows the pumping system to collect a high proportion of LNAPL in relation to the amount of groundwater. For best operation, the LNAPL thickness should be sufficient to completely cover the pump suction port. Figure 1-1 compares LNAPL and groundwater flows, when using a conventional dual-pump system vs. a conventional skimmer.

Drawdown pumping is a commercially available technology that can be easily implemented with conventional pumps in wells or trenches. System installation costs are low to moderate, but the cost per amount of LNAPL recovered varies greatly. It is difficult to predict the quantity of recoverable LNAPL using normal site characterization data such as LNAPL levels in wells or soil LNAPL concentration. The quantity recovered may be much higher or lower than initially estimated, resulting in lower or higher cost per unit volume recovered.

1.2 Bioslurping

Bioslurping combines vacuum-assisted LNAPL recovery with bioventing and SVE to simultaneously recover LNAPL and bioremediate the vadose zone. A bioslurper system withdraws free-phase LNAPL from the water table, relatively small amounts of groundwater, and soil gas/vapor in the same process stream using the air lift created by a single aboveground vacuum pump. Groundwater is then separated from the free product and is treated (when required) and discharged. Free product is recovered and can be recycled. Soil gas/vapor is treated (when required) and discharged.

1.2.1 Bioslurper Technology Description

Bioslurping can improve LNAPL recovery efficiency compared to other LNAPL recovery technologies. The bioslurper system may pull a vacuum of up to 25 ft of water on a recovery well to create the pressure gradient needed to force movement of LNAPL into the well (Figure 1-1). The system is operated to minimize drawdown in the water table, thus reducing the problem of free-product entrapment in soil.

Bioventing of the vadose zone soil is achieved by withdrawing soil gas via the recovery well. The slurping action of the system cycles between recovering liquid (free product and/or groundwater) and soil gas. The rate of soil-gas extraction is dependent on the rate of liquid recovery into the well. When free-product removal activities are complete, the bioslurper system can be converted easily to a conventional bioventing system to complete remediation of the vadose zone soil.

Preliminary data from short-term bioslurper tests conducted by Battelle for the U.S. Air Force Center for Environmental Excellence (AFCEE) and the Naval Facilities Engineering Service Center (NFESC) indicate that the LNAPL recovery rate achieved by bioslurping is up to six times greater than the rate achieved by skimming and as much as two times greater than the rate achieved by drawdown pumping. Mathematical models comparing drawdown to bioslurping have predicted that the free-product mass removal from the affected soil would be three times as fast when using bioslurping (Parker, 1996). In addition, the models indicated that groundwater recovery rates may be seven times greater when drawdown technology is used. Therefore, the disposal costs for groundwater could be less with bioslurping due to the lower groundwater recovery rates. Because the performance and process efficiency of these technologies depend heavily on the site characteristics, it is difficult to compare the costs for these technologies. Because the bioslurper system does appear to remove free product more rapidly than conventional pumping technologies, it is reasonable to assume that operations and maintenance (O&M) costs would be lower than for conventional technologies. A bioslurping implementation cost-estimating guide is included as Appendix A. The guide can be used to make reasonable cost estimates for the bioslurper installation, operations, and maintenance.

In summary, the preliminary analysis of the available field data indicates that bioslurping is a cost-competitive technology for LNAPL recovery with the added advantage of simultaneous vadose zone remediation. Like skimming and drawdown pumping, bioslurping would be less effective in tight (low-permeability) soils. Bioslurping is applicable at sites with a deep groundwater table (>25 ft), although adjustments to the system components, such as pump sizing, are required to increase the air lift needed to entrain LNAPL and water droplets. Prior to technology selection, the feasibility of LNAPL recovery must be evaluated for each technology based on site characterization data. If the evaluation indicates that bioslurping is the most practical, then the data required for the system design should be generated as discussed in Chapters 2.0, 3.0 and 4.0 of this book.

1.2.2 Principles and Theory of LNAPL Migration and Vacuum-Enhanced Recovery

1.2.2.1 LNAPL Migration

LNAPL generally enters the environment from leaks in fuel storage tanks or fuel transfer lines. Occasionally, LNAPL will enter the environment from surface spills; however, surface spills generally are not large enough to generate large-scale LNAPL plumes. The migration of LNAPL in the subsurface is dependent on the volume of LNAPL released, the depth to the groundwater area of infiltration, the time duration of the release, properties of the LNAPL, properties of the media, and subsurface flow conditions. When LNAPL is released at or near the surface, it migrates vertically through the vadose zone under gravitational and capillary forces. As the LNAPL passes through the vadose zone, much of it may be left in the pore spaces due to surface tension effects. Figure 1-2 displays the progression of an example LNAPL release.

If the volume of LNAPL released to the subsurface is great enough, some of the LNAPL will reach the saturated zone. Here the LNAPL will spread laterally along the capillary zone and may depress the water table. In the capillary fringe zone the fuel shares the pore spaces of the formation with soil gas and water. Figure 1-3 provides a conceptual distribution of air, LNAPL, and water near the water table. Below the water table, water occupies all the pore spaces (except when stable emulsions are present). Just above the water table, both LNAPL and water share the pore spaces.

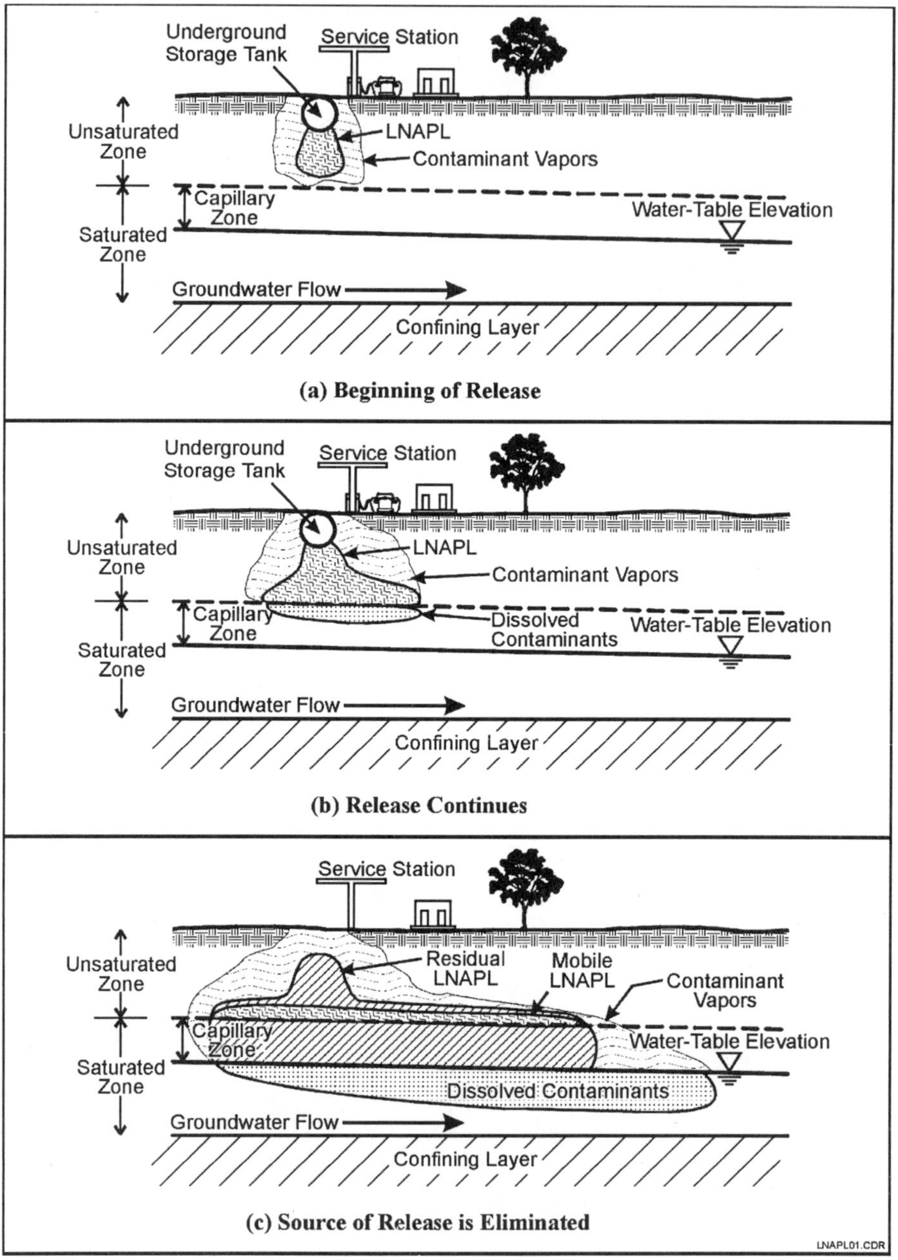

FIGURE 1-2. Progression of an example LNAPL release (Modified from U.S. EPA, 1990)

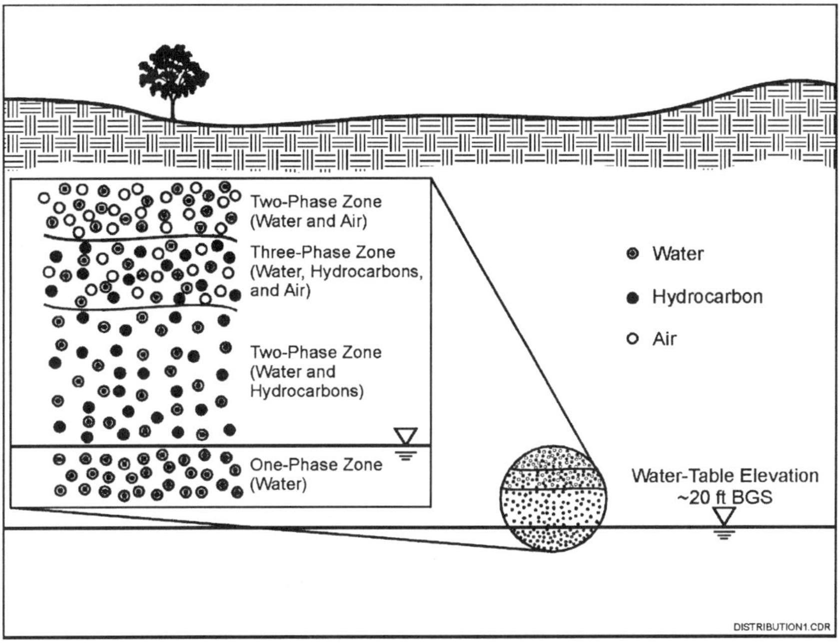

FIGURE 1-3. Distribution of air, hydrocarbon, and water in a porous medium and an observation well

This zone can be subdivided into a section where both the LNAPL and water are at positive pressure and an overlying section where the water is at a negative pressure (i.e., suction) and LNAPL is under a positive pressure. Water, LNAPL, and soil gas fill the pore spaces in the three-phase zone; and LNAPL in this zone exists under a negative pressure. Finally, if the source has been stopped, there may be a two-phase zone that contains water and soil gas.

The subsurface migration of LNAPL is controlled by the following factors:

- **Density (ρ).** Density is the mass of substance per unit volume and often is expressed as specific gravity. The specific gravity is the ratio of the mass of a volume of a substance at a given temperature to the mass of the same volume of water at a specified temperature. By definition, LNAPL is less dense than water, indicating that water will apply an upward buoyant force to LNAPL. A density difference as small as 1% can influence the interaction of two immiscible fluids in the subsurface (Mackay et al., 1985).

- **Viscosity (μ).** Viscosity measures the resistance of a fluid to flowing under the influence of an applied pressure gradient. Dynamic, or absolute, viscosity is measured in units of mass per unit length per unit time. A low viscosity (thin) fluid will move more rapidly through soil than a high viscosity (thick) fluid, assuming other factors (e.g., interfacial tension) are equal.

- **Water Solubility.** LNAPL exists as a separate phase in the presence of water because it is primarily a mixture of polar compounds that have low water solubility. However, LNAPL constituents range from slightly soluble (e.g., benzene) to nearly insoluble (e.g., *n*-dodecane). A few additives in some types of LNAPL (e.g., methyl-*tert*-butyl ether [MTBE] in gasoline) are very soluble in water.

- **Vapor Pressure (P).** Volatilization refers to mass transfer from liquid or soil to the gaseous phase. The vapor pressure of a compound is the main property affecting its tendency to volatilize. Volatile compounds at low concentration in an LNAPL mixture often contribute a high proportion to the vapor-phase contaminant concentrations. This behavior is described as Raoult's law, which says that the vapor pressure of a compound above a mixture of miscible liquids containing the compound is equal to the pure compound vapor multiplied by the mole fraction of the compound in the liquid.

- **Henry's Law Constant (K_H).** Compounds that have dissolved into the aqueous phase will evaporate across an air/water interface to establish an equilibrium concentration in the vapor. The vapor concentration of the compound is determined by the Henry's law constant. The Henry's law constant of low solubility hydrocarbon constituents of LNAPL can be estimated as the pure component vapor pressure divided by the pure component water solubility (Mackay and Shiu, 1981).

- **Sorption of Contaminants.** This is a complex process involving several different phenomena including coulomb forces, London-van der Waals forces, hydrogen bonding, ligand exchange, dipole-dipole forces, dipole-induced dipole forces, and hydrophobic forces. In the case of hydrocarbons, due to their nonpolar nature, sorption most often occurs through hydrophobic bonding to organic matter. Hydrophobic bonding often is an important factor influencing the fate of dissolved organic chemicals in the subsurface (DeVinny et al., 1990). The sorption coefficient of an organic compound in soil generally is empirically related to the organic content of the soil and the octanol/water partitioning coefficient of the compound.

Sorption isotherms generally follow one of three shapes: Langmuir, Freundlich, or linear. The Langmuir isotherm describes the sorbed contaminant concentration as increasing linearly with total contaminant concentration, then leveling off as the number of sites available for sorption are filled. This isotherm accurately describes the situation near the contaminant source where concentrations are high. The Freundlich isotherm assumes an infinite number of sorption sites, which would accurately describe an area some distance from the contaminant source where concentrations are dilute. The mathematical expression contains a chemical-specific coefficient that may alter the linearity of the isotherm. The linear isotherm is relatively simple and is valid for dissolved compounds at less than one-half of their solubility (Lyman et al., 1992). The linear isotherm is typically valid to describe hydrocarbon sorption.

The linear isotherm is expressed mathematically as:

$$C_s = K_d C_w \tag{1-1}$$

where: C_w = volumetric concentration of contaminant x in the aqueous phase ($g_x/L_{aqueous}$)
C_s = quantity of contaminant x sorbed to the soil (g_x/g_{soil})
K_d = sorption coefficient.

The sorption coefficient may be determined experimentally, estimated based on values published in the literature, or estimated using contaminant and soil properties. Nonpolar organic compounds sorbed on soil usually are held by hydrophobic interactions resulting in weak, nonspecific sorption forces. The sorption coefficient is a function of the hydrophobic character of the compound and the amount of organic matter present, and may be expressed as (Charbeneau and Weaver, 1992):

$$K_d = K_{oc} f_{oc} \tag{1-2}$$

where: K_d = sorption coefficient
K_{oc} = organic carbon partitioning coefficient
f_{oc} = fraction of organic carbon in the soil ($g_{organic\ carbon}/g_{soil}$).

The organic carbon partitioning coefficient often is estimated as being equal to the octanol/water partitioning coefficient (K_{oc}) (U.S. EPA, 1991). Hydrocarbons tend to become more hydrophobic with increasing carbon number, so larger hydrocarbons tend to sorb more strongly in comparison to smaller hydrocarbons.

- **Saturation (s).** Saturation is the fractional volume of the total pore volume occupied by that fluid. This value ranges from 0.0 to 1.0. In general, the permeability of a particular fluid or phase increases as the saturation of that fluid or phase increases.

- **Interfacial tension (σ).** Interfacial tension is the amount of work required to separate a unit area of one substance from that of another (Fetter, 1993), and determines the stability of the interface between the substances. Liquid interfacial tension is directly related to the capillary pressure across an LNAPL/water interface and is a factor controlling wettability. Higher surface tension typically increases the residual saturation level. Interfacial tension decreases as temperature increases and is affected by surfactants, pH, and dissolved gases.

- **Wettability.** Wettability refers to the relative affinity of a soil matrix for a fluid, and affects the in situ distribution of fluids. Wettability indicates the overall tendency of one fluid to spread over and adhere to a solid in the presence of another fluid with which it is immiscible. The wetting fluid will tend to coat the surface of the grains and occupy the smaller spaces of the porous medium, and the nonwetting fluid will tend to be located in the largest pore spaces. For aquifers, water generally is the wetting fluid (Fetter, 1993).

- **Capillary pressure (P_c), where $P_c = P_n - P_w$.** Capillary pressure is the differential pressure across the interface of wetting (w) and nonwetting (n) fluids. This pressure is a function of the size of the pores, the angle formed by the fluid surface in contact with the pore wall, and the interfacial tension at the surface of the wetting phase. As such, capillary pressure determines the pore size that can contain the interface. In the presence of water, LNAPL can enter a porous medium only when the capillary pressure of the largest pore is exceeded. Capillary pressure increases with decreasing pore size, decreasing moisture content, and increasing interfacial tension. Capillary pressure may be determined experimentally or by using the equations by Barnes and McWhorter (1995) presented in Section 1.2.2.2.

- **Residual saturation (S_r).** This is the saturation at which LNAPL becomes discontinuous and is immobilized by capillary forces under ambient groundwater flow conditions (Mercer and Cohen, 1990). Residual saturation results from capillary forces and depends on several factors, including the medium pore size distribution, wettability, fluid viscosity ratio and density ratio, interfacial tension, gravity/buoyancy forces, and hydraulic gradients.

- **Relative permeability (k_r).** Relative permeability is a parameter that represents the ratio of soil permeability to a particular fluid at a specific saturation level to its permeability at 100% saturation. When two immiscible fluids are present in the pore spaces of the medium, part of the pore space will be filled with one fluid and the remainder will be filled with the other fluid. Relative permeability is different for wetting and nonwetting fluids. Figure 1-4 presents two-phase relative permeability curves for both wetting and nonwetting fluids. Relative permeability in a three-phase system generally is determined by laboratory testing of the contaminated medium; however, such measurements are difficult and expensive. Theoretical models have been developed to characterize three-phase relative permeability (Stone, 1973). Faust (1985) presented the relative permeability of nonaqueous-phase liquid (NAPL), k_{rn}, in a three-phase system as:

$$k_{rn} = k^*_{rnw}\left[\left(\frac{k_{rnw}}{k^*_{rnw}} + k_{rm}\right)\left(\frac{k_{rna}}{k^*_{rnw}} + k_{ra}\right) - \left(k_{rw} + k_{ra}\right)\right] \quad (1\text{-}3)$$

where: k^*_{rnw} = the relative permeability of the nonaqueous phase at the residual saturation of water in a water/nonaqueous system.

k_{rnw} = the relative permeability of the nonaqueous-phase system as a function of water saturation.

k_{rna} = the relative permeability of the nonaqueous phase in an air nonaqueous-phase system as a function of air saturation.

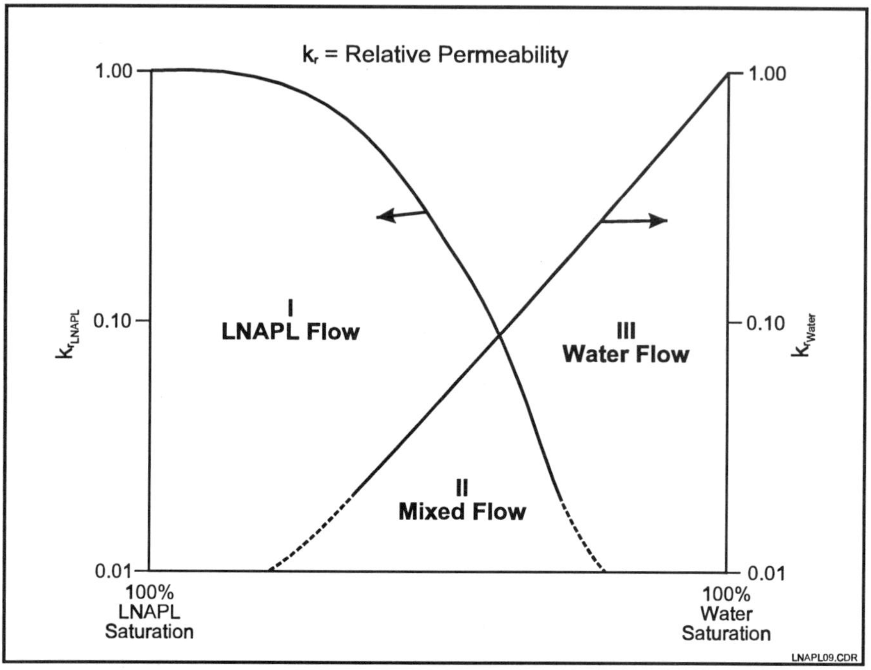

FIGURE 1-4. Hypothetical relative permeability curves for water and LNAPL in a porous medium (Modified from U.S. EPA, 1995)

1.2.2.2 Vacuum-Enhanced Recovery

Vacuum-enhanced recovery is a common groundwater pumping technique used for construction dewatering (Powers, 1981). Vacuum-enhanced pumping involves the application of a negative pressure to a well point system to increase the flowrate of groundwater and soil gas into the well. Recently, vacuum-enhanced pumping has been applied to the groundwater remediation pump-and-treat and LNAPL recovery systems. Increased groundwater extraction rates and increased LNAPL recovery have been documented by Blake and Gates (1986) and Parker (1996). Applying vacuum-enhanced pumping to hydrocarbon-contaminated sites also facilitates the following activities (Blake et al., 1990):

- Increased liquid recovery and hydraulic gradient control
- Vapor and residual hydrocarbon removal
- Combined vapor recovery and soil-gas pressure control.

Reisinger et al. (1993) reported a 47% increase in groundwater extraction as a result of vacuum-enhanced pumping.

Two important factors influence the movement of LNAPL into a recovery well: (1) the hydraulic gradient (or hydraulic head difference between the well and the surrounding strata), and (2) the relative (LNAPL) permeability of the aquifer (or the rate at which LNAPL moves through the aquifer). Vacuum-enhanced recovery improves recovery rates by increasing the hydraulic gradient and aquifer transmissivity. Conventional dual-pump free-product recovery systems increase the hydraulic gradient to a well by setting a pump below the water table to establish a cone of depression in the water table around the well. Free product then flows down the gradient diagonally into the well under gravitational forces to be recovered by a second LNAPL extraction pump. Vacuum-enhanced pumping systems use the same concept, except that the cone of depression actually is a cone of reduced pressure around the well. Parker (1996) compares vacuum-enhanced extraction with drawdown extraction and presents the relationship between the gradient produced by vacuum-enhanced and drawdown-LNAPL recovery systems with the term *effective drawdown*.

$$\Delta H = \Delta Z + P_a \tag{1-4}$$

where: ΔH = total water piezometric drawdown in the extraction well
ΔZ = the amount of drawdown in an extraction well (feet of water)
P_a = vacuum in the extraction well (feet of water).

Fluids flow horizontally across the pressure-induced gradient from higher pressure outside the well to lower pressure inside the well. The transmissivity of the saturated zone is an intrinsic characteristic of an aquifer and is a function of the hydraulic conductivity and the saturated thickness of the aquifer. Vacuum-enhanced pumping increases transmissivity by promoting flow along more-permeable horizontal flow lines and by decreasing the local pressure above the aquifer to effectively increase the saturated thickness of the aquifer. In addition, vacuum-enhanced pumping promotes continuity in the LNAPL phase (i.e., lower capillary pressure, which results in fewer air pockets in the capillary fringe). The combined effect of increased hydraulic gradient and aquifer transmissivity results in an enhanced liquid recovery rate.

Barnes and McWhorter (1995) present the requirements for LNAPL flow toward the well from a three-phase zone containing air, water, and LNAPL. First, they define the capillary pressure in a three-phase system as:

$$h_c^{ao} = h_a - h_o \tag{1-5}$$

and

$$h_c^{ow} = h_o - h_w \tag{1-6}$$

where: h_c^{ao} = air/LNAPL capillary pressure
h_c^{ow} = LNAPL/water capillary pressure
h_a = air pressure
h_o = hydrocarbon pressure
h_w = water pressure.

The condition of each of the phases outside the effect of the bioslurper system, or the far field is defined next. It is assumed that the soil gas is in direct communication with the atmosphere, so h_a is equal to zero-gauge pressure. The pressure of the LNAPL in the far field and where $h_a = 0$ is defined as:

$$h_o = \frac{\rho_o}{\rho_w}(z - T), \qquad z \geq T \tag{1-7}$$

and the water pressure is defined as:

$$h_w = \frac{\rho_o}{\rho_w} T - z, \qquad z \geq T \tag{1-8}$$

where: ρ_o = density of the LNAPL
ρ_w = density of water
z = elevation from the LNAPL/water interface in the observation well
T = thickness of LNAPL in the observation well.

A vacuum that is produced in a bioslurper extraction well reduces the prepumping air/oil and oil/water capillary pressures in the formation around the well. The vacuum generated at the extraction well is treated as drawdown of air pressure. LNAPL will flow to the well when the air/LNAPL capillary pressure at the well is zero. This condition occurs when the following equation is satisfied:

$$h_c^{ao}(r = \infty) \geq \Delta h_a(r_w) \tag{1-9}$$

where: $h_c^{ao}(r = \infty)$ = air/LNAPL capillary pressure in the far field
$\Delta h_a(r_w)$ = air pressure at the well.

The preceding equations indicate the minimum vacuum required to enhance the migration of LNAPL to the well; however, these equations do not provide the flow velocity of fuel in the formation. It should be stated again that Equations 1-7 through 1-9 are for LNAPL in the three-phase zone where LNAPL exists under negative pressure. In the two-phase zone (LNAPL and water), the LNAPL is under positive pressure. Therefore, if the minimum vacuum is achieved for the three-phase zone, LNAPL in the two-phase zone also will migrate toward the extraction well. The flow velocity of LNAPL at the minimum vacuum required to induce LNAPL flow may be so slow that remediation of the site in a practical time frame is impossible. The following equation may be used to estimate the Darcy velocities (q) in p-phase (p = w, o, a for water, oil, and soil gas):

$$q_{pi} = -K_{pij}\left(\frac{\partial \psi_p}{\partial x_j} + \rho_{rp} u_j\right) \qquad (1\text{-}10)$$

where: K_{pij} = is the p-phase conductivity tensor
i, j = directional indices (i, j = 1, 2, 3) with repeated values indicating summation in tensor notation, x_i (or x_j) is the *i*th (or *j*th) Cartesian coordinate
ψ_p = $(P_p/g\rho^*_w)$ = the water equivalent pressure head of phase p
P_p = the p-phase pressure
g = gravitational acceleration
ρ^*_w = the density of pure water
ρ_{rp} = (ρ_p/ρ^*_w) is the p-phase specific gravity
ρ_p = the p-phase density
u_j = $(\partial z/\partial x_j)$ is a unit gravitational vector measured as a positive, upwards.

As mentioned previously, Equation 1-10 can provide a general velocity for the fuel. However, the velocity of the fuel during extraction will change due to the saturation and relative permeability of each of the phases. Thus, the LNAPL velocity will change over time and location relative to the well.

Several mathematical models have been developed to determine the flow velocity of LNAPL in the subsurface. These models account for changes in the saturation and relative permeability of the phases over time and location relative to the extraction well. An example of such a model is the MOVER code (Draper Aden Environmental Modeling, Inc. [DAEM], 1997), which calculates the velocities of the LNAPL, groundwater, and soil gas during bioslurping. Multiple-phase models also may be able to predict the time required for site remediation and total volume of recovered LNAPL.

Suction lift might appear to be a limitation to the application of vacuum-enhanced dewatering. In theory, the maximum suction lift attainable with an extremely efficient vacuum pump is approximately 25 ft, depending on elevation (Powers, 1981). In practice, however, lifts greater than the theoretical maximum can be attained when the extracted fluid is not only water, but a mixture of soil gas and groundwater (Powers, 1981). Battelle has achieved liquid extraction from a depth of 210 ft at one location. A mixture of soil gas and water has a specific gravity less than 1.0 and, therefore, can be lifted higher than a standard water column. When LNAPL (specific gravity <1.0) is extracted with the soil gas and groundwater, there is a greater increase in suction lift. Another phenomenon that can help in achieving greater increase than the theoretical suction lift is liquid entrainment or entrapment. Liquid entrainment occurs when the primary extraction fluid is soil gas, rather than a liquid. At high velocities, extracted soil gas can entrain water droplets and carry them to the surface via slug flow at high liquid extraction rates.

1.2.3 Bioventing Component of Bioslurping

Bioventing is the process of aerating vadose zone soil, thereby stimulating soil-indigenous microorganisms to aerobically metabolize petroleum hydrocarbons in the soil. Application of bioventing has been tested extensively by Battelle at petroleum hydrocarbon-contaminated sites. Bioslurping is similar in design to soil venting (a.k.a. soil vacuum extraction, soil-gas extraction, or in situ soil stripping). The significant difference is that soil venting is designed and operated to maximize volatilization of low-molecular-weight compounds, although some biodegradation also occurs. In contrast, bioventing is used to maximize biodegradation of aerobically biodegradable compounds, regardless of volatility. Although all of these technologies involve venting of air through the vadose zone, the differences in objectives result in significantly different designs and operations of the remedial systems.

Petroleum distillate fuel hydrocarbons, such as JP-5 and JP-8 jet fuel, are biodegradable as long as naturally occurring microorganisms are provided with an adequate supply of oxygen and basic nutrients (Atlas, 1981). Natural biodegradation does occur at many contaminated sites and eventually mineralizes most fuel contaminants. However, the degradation process is dependent on the natural oxygen diffusion rate at the site (Ostendorf and Kampbell, 1989), which frequently is too slow in most contaminated soils to promote effective biodegradation. At such sites, acceleration of the oxygen transport process via bioventing may prove to be the most effective way to enhance bioremediation.

The main features of bioventing technology include the following:

- Optimizing airflow to minimize volatilization while maintaining aerobic conditions for biodegradation
- Monitoring local soil-gas conditions to ensure that aerobic conditions exist (not just monitoring vent gas composition)
- Conducting in situ respiration tests that provide for the effective measurement of continued contaminant biodegradation
- Manipulating the water table as required for air/contaminant contact.

1.2.4 Soil Vapor Extraction Component of Bioslurping

At sites contaminated with fuels possessing relatively high volatilities (e.g., JP-4 and gasoline), the SVE component of the bioslurper becomes an important process for contaminant removal. For example, removal of contaminants in the vapor phase was as high as 980 lb/day at a gasoline-contaminated bioslurper site at Eaker Air Force Base (AFB) (Battelle, 1996). Note that although a portion of the released vapor-phase LNAPL is from the extracted soil gas, the majority may be from the volatilization of extracted free product as it passes through the bioslurper system. Often, the fraction of the vapor-phase contaminants from volatilization may be estimated by subtracting the average contaminant concentration in the soil gas from the total contaminant concentration in the stack gas. However, the bioslurper system draws atmospheric air into the subsurface when the vacuum is generated and the total petroleum hydrocarbon (TPH) concentration in the soil gas may be diluted. Therefore, dilution of the soil-gas TPH concentrations must be considered when estimating the fraction of contaminants in the vapor stream due to volatilization. If the oxygen concentrations in the gas remain constant throughout the study, significant dilution of the soil gas probably has not occurred.

ENSR (1995) summarizes the factors that affect the mass of contaminant removed from the subsurface in the vapor phase:

- Contaminant transport and removal

- Contaminant properties
  - — Vapor pressure
  - — Contaminant concentrations in the soil gas
  - — Henry's law constant for a contaminant
  - — Boiling point of a contaminant
  - — Soil adsorption coefficient (K_d)
  - — Octanol/water partitioning coefficient (K_{ow})
  - — Kinetics of soil adsorption/desorption
  - — Solubility of a contaminant

- Soil Properties
  - — Grain-size distribution
  - — Porosity
  - — Liquid saturation
  - — Fraction of wetting and nonwetting fluids
  - — Residual water saturation (S_r)
  - — Residual LNAPL saturation (S_{ro})
  - — Capillary pressure between wetting and nonwetting liquids (P_c)
  - — Moisture retention curves
  - — Intrinsic permeability (k)
  - — Hydraulic conductivity (K)
  - — Air permeability (k_a)
  - — Peclet number (relates mass transfer by advection to mass transfer by diffusion)
  - — Humidity

Typically, SVE systems extract soil gas at relatively high rates with the vacuum produced by a liquid ring pump (LRP) or a high-efficiency blower, such as a rotary lobe. However, the SVE rate for a bioslurper system may be much lower because the primary objective of bioslurping is to recover LNAPL and to extract soil gas at a rate sufficient to entrain liquid flow in the bioslurper extraction tube.

1.3 References

Atlas, R.M. 1981. "Microbial Degradation of Petroleum Hydrocarbons: An Environmental Perspective." *Microbial. Rev.*, *45*:180-209.

Baker, R. S. 1995. "One-, Two-, and Three-Phase Flow During Free-Product Recovery." In R.E. Hinchee, J.A. Kittel, and H.J. Reisinger (Eds.), *Applied Bioremediation of Petroleum Hydrocarbons*, pp. 349-359. Battelle Press, Columbus, OH.

Barnes, D.L., and D.B. McWhorter. 1995. "Mechanics of Vacuum-Enhanced Recovery of Hydrocarbons." In R.E. Hinchee, J.A. Kittel, and H.J. Reisinger (Eds.), *Applied Bioremediation of Petroleum Hydrocarbons*, pp. 361-370. Battelle Press, Columbus, OH.

Battelle. 1996. *Site-Specific Report for Free-Product Recovery Testing at Eaker AFB, AR.* Prepared for the Air Force Center for Environmental Excellence, San Antonio, TX.

Blake, S.B., and M.M. Gates. 1986. "Vacuum Enhanced Hydrocarbon Recovery: A Case Study." In: *Proceedings of the NWWA/API Conference on Petroleum Hydrocarbons and Organic Chemicals in Ground Water: Prevention, Detection, and Restoration.* Water Well Journal Publishing Company, Dublin , OH. pp. 709-754.

Blake, S.B., B. Hockman, and M. Martin. 1990. "Applications of Vacuum Dewatering Techniques to Hydrocarbon Remediation." In: *Proceedings of Petroleum Hydrocarbons and Organic Chemicals in Ground Water: Prevention, Detection, and Restoration.* Water Well Journal Publishing Company, Dublin , OH. pp. 211-225.

Charbeneau, R.J., and J.W. Weaver. 1992. "Modeling Contaminant Transport through Subsurface Systems." *Journal of Hazardous Materials, 32*: 293-311.

DAEM, see Draper Aden Environmental Modeling, Inc.

DeVinney, J.S., L.G. Everett, J.C.S. Lu, and R.L. Stollar. 1990. *Subsurface Migration of Hazardous Wastes.* Van Nostrand Reinhold, New York, NY.

Draper Aden Environmental Modeling, Inc. 1997. OILVOL: Oil Volume Estimator. Technical Documentation and User's Guide. Blacksburg, VA. 11 p.

ENSR. 1995. *Soil Vapor Extraction and Bioventing.* EM1110-1-4001. Prepared by ENSR under contract to U.S. Army Corps of Engineers. November.

Faust, C.R. 1985. "Transport of Immiscible Fluids Within and Below the Unsaturated Zone; a Numerical Model." *Water Resources Research, 21*(4): 587-96.

Fetter, C.W. 1993. *Contaminant Hydrogeology.* Macmillan Publishing Company, New York, NY.

Leeson, A., J.A. Kittel, R.E. Hinchee, R.N. Miller, P.E. Haas, and R.E Hoeppel. 1995. "Test Plan and Technical Protocol for Bioslurping." In R. E. Hinchee, J. A. Kittel, and H.J. Reisinger (Eds.), *Applied Bioremediation of Petroleum Hydrocarbons*, pp. 335-347. Battelle Press, Columbus, OH.

Lyman, W.J., P.J. Reidy, and B. Levy. 1992. *Mobility and Degradation of Organic Contaminants in Subsurface Environments.* C.K. Smoley, Chelsea, MI.

Mackay, D., and W.Y. Shiu. 1981. "A Critical Review of Henry's Law Constants for Chemicals of Environmental Interest." *Journal of Physical and Chemical Reference Data, 10*(4): 1175-1199.

Mackay, D., P.V. Roberts, and J.A. Cherry. 1985. "Transport of Organic Contaminants in Groundwater." *Environmental Science and Technology, 19*(5): 384-392.

Mercer, J.W., and R.M. Cohen, 1990. "A Review of Immiscible Fluids in the Subsurface: Properties, Models, Characterization and Remediation." *Journal of Contaminant Hydrology, 6*: 107-163.

Ostendorf, D.W., and D.H. Kampbell. 1989. "Vertical Profiles and Near Surface Traps for Field Measurement of Volatile Pollution in the Subsurface Environment." In: *Proceedings of the NWWA Conference on New Techniques for Quantifying the Physical and Chemical Properties of Heterogeneous Aquifers*, Dallas, Texas. ISBN 389432-009-5. Westarp Wiss. Essen, Germany, pp. 475-485.

Parker, J.C. 1996. "Evaluating the Effectiveness of Product Recovery, Bioventing, and Bioslurping Systems." *Environmental Systems and Technology Inc. News, 4*(Winter 95): 4-6. Blacksburg, VA.

Powers, J.P. 1981. *Construction Dewatering*. John Wiley & Sons, New York, NY. pp. 290-298.

Reisinger, H.J., P. Hubbard, S.A. Mountain, and C.W. Brigham. 1993. "Integrated Site Remediation System Using High Vacuum Application to Address Ground-Water Extraction, Soil Venting and *In Situ* Biodegradation." Paper presented at U.S. EPA Groundwater Remediation/Stabilization Conference, held in Atlanta, Georgia. December 1-3, 1993.

Stone, H.L. 1973. "Estimation of Three-Phase Relative Permeability and Residual Oil Data." *Journal of Canadian Petroleum Technology, 12*(4): 53-61.

U.S. Environmental Protection Agency. 1990. *Assessing UST Corrective Action Technologies: Early Screening of Cleanup Technologies for the Saturated Zone*. EPA/600/2-90/027. Risk Reduction Engineering Laboratory, Cincinnati, OH.

U.S. Environmental Protection Agency. 1991. *Soil Vapor Extraction Technology: Reference Handbook*. EPA/540/2-91/003. Office of Research and Development, Risk Reduction Engineering Laboratory, Cincinnati, OH.

U.S. Environmental Protection Agency. 1995. *Ground Water Issue: Light Nonaqueous Phase Liquids*. EPA/540/S-95/500. Office of Solid Waste and Emergency Response, Washington, DC.

U.S. Environmental Protection Agency. 1997. *Use of Monitored Natural Attenuation at Superfund, RCRA Corrective Action, and Underground Storage Tank Sites*. EPA/9200.4-17. Office of Solid Waste and Emergency Response, Washington, DC.

U.S. EPA, see U.S. Environmental Protection Agency.

2 Site Characterization and Evaluation

Field tests are required to evaluate the feasibility of the bioslurping technology and to generate the required data to design and install a full-scale bioslurping remediation system. The first step of the bioslurper feasibility evaluation involves reviewing the site characterization data. Based on these data, a pilot test should be performed at a location representative of the site characteristics and contamination. If the geologic and physical characteristics of the subsurface vary significantly at the site, pilot tests at more than one location may be required.

2.1 Site Characterization

It is important to conduct a thorough review of site characterization data, including information sources describing when the release of LNAPL occurred, the quantity and type of the LNAPL released, measured LNAPL thickness in monitoring wells located in the area of concern, petroleum hydrocarbon levels in soils, areas/extent of contamination, and the site geology and hydrogeology. Most of this information is expected to be in the initial assessment and confirmation studies, site characterization reports, and remedial investigation/feasibility studies. If the available information is limited, a site characterization program may have to be implemented to obtain the above data.

2.1.1 LNAPL Release History

A detailed history of the LNAPL release(s) should be obtained as part of the site characterization. Generally, the history of the release can be found in site characterization documents. If such documents have not been prepared, the remedial project manager (RPM) or the site personnel may be able to provide information regarding the history of the LNAPL release. The following information should be gathered when constructing the release history:

- The dates when the release(s) occurred
- The locations of the release(s)
- The volume of LNAPL discharged during each release
- The estimated time duration of each release
- The type of LNAPL involved in each release
- Whether (and when) the source of the release was removed
- The mass of contaminants recovered after the release (to date).

2.1.2 Areas/Extent of Characteristics of Contaminants

Proper delineation of the extent of contamination is important for selection of the pilot test location and for the full-scale implementation of an LNAPL-recovery system. The thickness and depth of the LNAPL layer are determined through interface measurements. Baildown testing provides information regarding the mobility of the LNAPL under passive conditions. A soil-gas survey may provide the location of the most contaminated areas of the LNAPL plume. Along with the soil-gas survey, in situ respiration testing will provide the potential for biodegradation of the contaminants in the vadose zone at the site.

2.1.2.1 Interface Measurements

The depth to groundwater and apparent thickness of LNAPL in monitoring wells can be measured with an oil/water interface probe, which distinguishes between polar and nonpolar fluids in the well. Such measurements should be conducted several times per year to determine if temporal variations occur in the interface readings.

It has been demonstrated that the thickness of LNAPL in a monitoring well is greater than the thickness of LNAPL in the formation (de Pastrovich et al., 1979; Lenhard and Parker, 1990), and that the apparent thickness in a monitoring well can be from 2 to 10 times greater than the actual LNAPL thickness in the formation (Mercer and Cohen, 1990). Lenhard and Parker (1990) developed an equation to estimate the actual LNAPL thickness in the formation using the apparent LNAPL thickness in the monitoring well (see Equation 2-1).

$$D_o = \frac{\rho_{ro}\beta_{ao}H_o}{\beta_{ao}\rho_{ro} - \beta_{ow}(1-\rho_{ro})} \quad (2\text{-}1)$$

where: D_o = actual thickness of LNAPL in the formation
H_o = apparent LNAPL thickness in the well
ρ_{ro} = density of LNAPL
β_{ao} = air/oil scaling factor = σ_{aw}/σ_{ao}
β_{ow} = oil/water scaling factor = σ_{aw}/σ_{ow}.

2.1.2.2 Baildown Testing

After the depth to groundwater and the initial LNAPL thickness have been determined, the rate of LNAPL recovery from the monitoring wells may be determined via baildown testing. In these tests, a clean Teflon™ bottom-filling bailer is lowered into each well to collect any floating LNAPL (LNAPL also may be removed from the well using a skimmer pump or a peristaltic pump). The collected LNAPL is poured into a graduated cylinder to determine its volume. Efforts should be made to minimize the volume of water removed from the well. Bailing should continue until all LNAPL has been removed from the well and the filter pack surrounding the well. Often, it is difficult to determine if all of the free product has been removed from the well and its annulus. As such, the same approach that is used for developing a well may be used to ensure that the free product has been removed from the sand pack. (When a well is developed for sampling, at least three borehole volumes of groundwater are removed from the well before sampling to ensure that the water to be sampled is coming from the formation.) It is recommended that at least two borehole volumes of LNAPL be removed from the well prior to initiating the recovery portion of the test.

After LNAPL removal, the LNAPL thickness in the well is monitored periodically using the oil/water interface probe to determine the rate of LNAPL recovery. Measurements may be taken every hour for 2 hours, then every 2 to 4 hours for a maximum of 24 hours. Measurements can be made more frequently if LNAPL recovery is rapid or less frequently if recovery is very slow. Data should be recorded on a baildown test record sheet (Figure 2-1). It is recommended that several baildown tests be performed throughout the year to monitor the change in the passive LNAPL recovery rate due to variations in the level of the water table, the soil moisture content, and the initial (prebailing) apparent thickness of LNAPL.

Site: ____

Well Identification: ____________________

Well Diameter (OD/ID): ________________

Date at Start of Test: ________ Sampler's Initials: _____

Time at Start of Test: ________________

Initial Readings

Depth to Groundwater (ft)	Depth to LNAPL (ft)	LNAPL Thickness (ft)	Total Volume Bailed (L)

Test Data

Sample Collection Time	Depth to Groundwater (ft)	Depth to LNAPL (ft)	LNAPL Thickness (ft)

FIGURE 2-1. Baildown test record sheet

2.1.2.3 Soil-Gas Survey

When existing monitoring wells are lacking, a soil-gas survey should be conducted to aid in locating a suitable location for installing a bioslurper well and soil-gas monitoring points. Ideally, the bioslurper well and soil-gas monitoring points should be located in soils containing measurable hydrocarbon contamination and where the oxygen level is depleted and the carbon dioxide level is elevated.

Soil-gas sampling can be conducted using small-diameter (e.g., 1-inch outside diameter [OD]) stainless steel probes with a slotted well point assembly. Hand-driven gas probes are used primarily at sites with a relatively shallow water table or where soils are penetrable to a depth of within 5 ft of the water table. The maximum depth for hand-driven probes typically is 10 to 15 ft, depending on the soil texture. As expected, penetration of soil-gas probes in dense silts and clays is less than in unconsolidated sands. At a given location, the probe should be emplaced (usually driven manually or with a power hammer) to a depth predetermined by a review of the site characterization/contamination documents. Soil gas at this depth should be analyzed with field instruments for oxygen, carbon dioxide, and TPH. Section 2.1.2.4 discusses the suitable equipment used for monitoring soil gas. The probe may then be driven deeper for additional soil-gas measurements. For a site with a depth to groundwater of 9 ft, soil gas should be measured at depths of 2.5, 5, and 7.5 ft.

In general, sites that contain low oxygen levels (i.e., 0 to 2%), high carbon dioxide levels (i.e., 5 to 20% depending on soil types), and relatively high TPH levels would be suitable candidate locations to be selected for bioslurper pilot testing. The TPH concentration in the soil gas is dependent on the type of fuel contaminating the site. With high-volatility fuels (gasoline, JP-4), a relatively high TPH concentration is >10,000 parts per million by volume (ppmv), whereas with low-volatility fuels (diesel, JP-5), a relatively high TPH concentration may be >1,000 ppmv. (An uncontaminated site also may be located for monitoring background respiration of natural organic matter and inorganic sources of carbon dioxide. Typical oxygen and carbon dioxide levels at an uncontaminated site are 15 to 20% and 1 to 5%, respectively; the TPH content generally is below 100 ppmv.)

Prior to sampling, soil-gas probes should be purged with a sampling pump. To ensure adequate purging, soil-gas concentrations should be monitored until they become stabilized. In general, removal of 1 to 2 L of soil gas is sufficient to purge the probes. This may not always be possible, particularly when soil-gas samples at shallow depths are collected allowing atmospheric air to be drawn into the probe, producing false readings. Figure 2-2 shows a typical setup for monitoring soil gas.

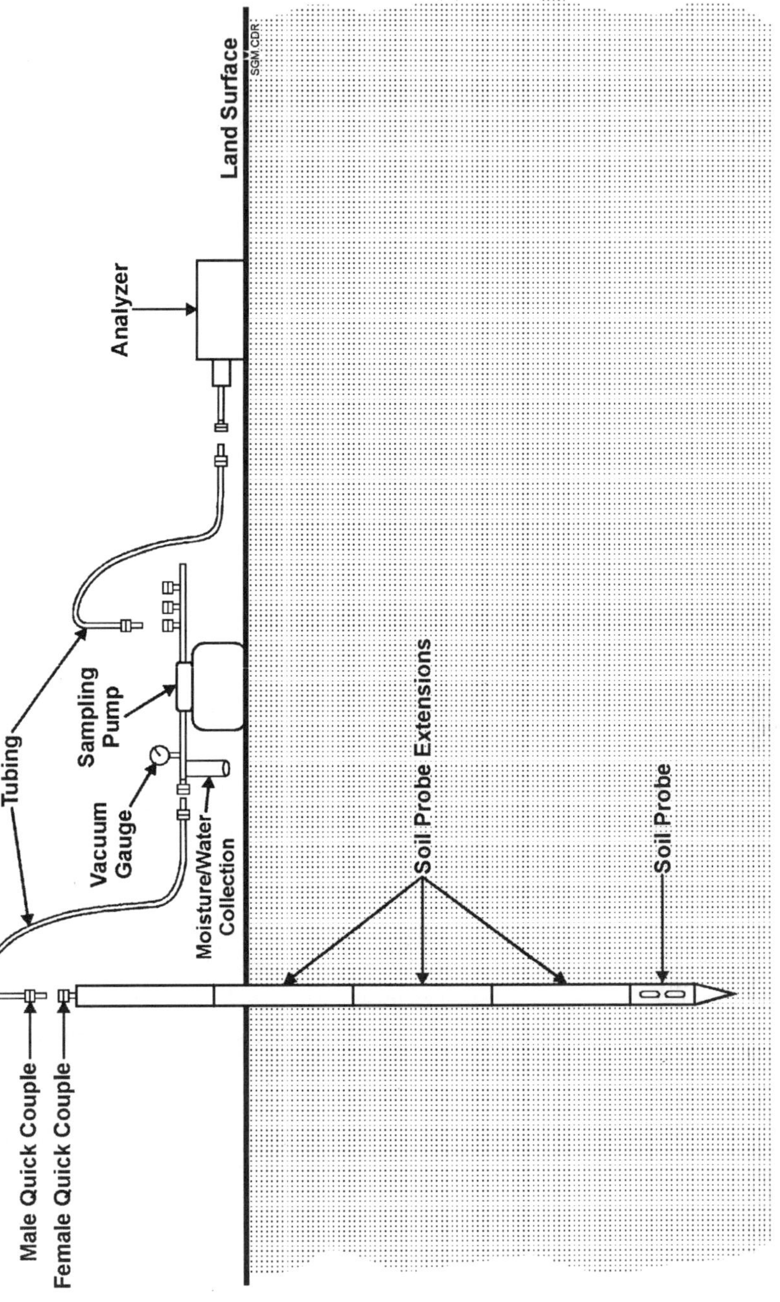

FIGURE 2-2. Typical setup for monitoring soil gas

2.1.2.4 In Situ Respiration Testing

Prior to performing any bioremediation testing, a baseline in situ respiration test should be completed. Leeson and Hinchee (1997) suggest that in situ respiration tests can be used to gauge the degree of bioremediation that has occurred throughout the remedial project. Monitoring the progress of site remediation in this manner reduces the number of soil samples needed to confirm site remediation results.

Air containing 1 to 2% helium is injected into soil-gas monitoring points for 24 hours to fully aerate the soil. The soil gas then will be measured for oxygen, carbon dioxide, and total TPH. Soil gas may be extracted from the contaminated area with a soil-gas sampling pump system similar to that shown in Figure 2-1 or using the soil-gas monitoring system discussed previously. Typically, the soil gas is measured with field instruments at 2, 4, 6, and 8 hours after startup, and then every 4 to 12 hours, depending on the rate at which oxygen is utilized. If oxygen uptake is rapid, more frequent monitoring will be required. If it is slow, less frequent readings may be acceptable. Soil-gas sampling for in situ respiration testing generally lasts for two days. The temperature of the soil before air injection and after the in situ respiration test should be recorded. If the oxygen depletion rate is very rapid (3 to 5% per hour), with little increase in the carbon dioxide content, some of the oxygen may be utilized to satisfy an immediate chemical oxygen demand in the soil such as caused by high reduced iron concentrations. If this situation is suspected, additional aeration may be required to satisfy this demand prior to conducting the test.

At shallow monitoring points, there is a risk of pulling in atmospheric air during purging and sampling. Also, excessive purging and sampling may result in erroneous readings. There is no benefit in oversampling. When sampling shallow points, care should be taken to minimize the volume of air extracted. In these cases, a low-flow extraction pump operating at 2 to 4 ft^3 per hour may be used. Field judgment is required at each site in determining the sampling frequency. Table 2-1 provides a summary of the various parameters that will be measured and the field equipment that can be used to monitor each parameter. The in situ respiration test can be terminated when the oxygen level is about 5%, or after two days of sampling.

2.1.2.5 LNAPL Plume Delineation and Volume Estimation

The LNAPL plume can be delineated using the available in-well air/oil and oil/water interface measurements. The accurate delineation of any contaminant plume is dependent on the scale of the site and the number of monitoring wells or piezometers. There is no general guideline defining the

TABLE 2-1. Parameters to be measured for the in situ respiration test

Parameter/ Media	Suggested Method	Suggested Frequency	Instrument Sensitivity (Accuracy)
Carbon dioxide/soil gas	CO_2 field meter, infrared absorption method (0 to 5% and 0 to 25% carbon dioxide)	Initial soil gas sample before pumping air, immediately after pump shutoff, every 2 hours for the first 8 hours, and then every 8 to 10 hours	0.2%
Oxygen/soil gas	O_2 field meter, electrochemical cell method (0 to 21% oxygen)	Same as above	0.5%
Total hydro-carbons/soil gas	Hydrocarbon detector	Initial soil gas sample before pumping air, then same as above if practical	1 ppm
Helium (optional)	Helium detector	Same as for carbon dioxide	0.01%
Pressure	Pressure gauge (0 to 30 psia)	Reading taken during air injection	0.5 psia
Flowrate/air	Flowmeter	Reading taken during air injection	cfh

cfh = cubic feet per hour.
ppm = parts per million.
psia = pounds per square inch absolute.

number of wells needed to define a plume, because every site is unique and the size of the site is an important factor in determining data needs. Contour maps of LNAPL thicknesses can be generated based on the interface measurements made in the field. Caution must be exercised, however, because LNAPL thickness measurements taken inside a well do not reflect the actual thickness of LNAPL in the formation.

A more accurate means of defining the extent of a LNAPL plume and estimating the total LNAPL volume at a site involves the determination of specific volumes of LNAPL across the area of interest. A complete discussion of specific oil volumes is provided in Farr et al. (1990) and in Lenhard and Parker (1990) (see Section 2.1.2.1 for a summary). Oil volume estimates can be calculated based on air/oil and oil/water level measurements made in the field using a computer program, such as OILVOL (DAEM, 1997). OILVOL is a computer program for estimating free hydrocarbon volume in the soil following a petroleum release. The elevations of the air/LNAPL interface and the LNAPL/water table measured in a network of monitoring wells are used to compute free hydrocarbon volume at the time fluid level observations are made.

OILVOL divides the problem domain into equally spaced rectangular grids. The difference between fluid-level elevations gives the free-product thickness at the monitoring well locations. These thicknesses are krigged to obtain the product thickness at each node in the rectangular grid of the problem domain. Using the three-phase constitutive relationship between phase saturation and pressure, the vertical distribution of water and LNAPL saturation is computed. Integration of the oil saturation with depth gives the specific oil volume (volume of free LNAPL per unit surface area). The specific oil volume at each grid node is multiplied by the respective nodal surface area to compute the volume of free hydrocarbon at the node, and these are summed to obtain the total free-product volume in the soil. The program is based on principles defined by Lenhard and Parker (1990).

The quality of the LNAPL-volume estimate is dependent on the quality of field and laboratory measurements for fluid levels, soil properties, and fluid properties. Care must be taken when the field and laboratory measurements are performed. In addition, LNAPL may exist as small pools as opposed to one large plume. It is important that reliable data are used when the LNAPL volume is estimated. Erroneous LNAPL volume estimates calculated by the computer program lead to false expectations of the recoverable LNAPL volume.

Workers have used laser-induced fluorescence (LIF) to determine the vertical and horizontal distribution of LNAPL in the subsurface. The site characterization units attach LIF probes to direct push penetrometers. An LIF probe emits ultraviolet light through a sapphire window near the tip of the penetrometer rod. When exposed to ultraviolet light, polycyclic aromatic hydrocarbons (PAHs) produce a fluorescent response. The intensity and wavelength of the fluorescent response is detected and measured by the LIF system. As the probe is pushed through the soil, the system generates fluorescent intensity and wavelength of the maximum intensity versus depth to provide the vertical distribution of hydrocarbons.

LIF units are a semiquantitative analysis technique. The resulting fluorescent intensity is a measure of the response to the input energy. The greater the input energy, the greater the energy of the response. In addition, the response increases linearly with the PAH concentration. However, the intensity of the response is dependent on the type of PAH present in the subsurface. LIF units are more effective for use with hydrocarbons containing relatively high concentrations of PAHs with three or more rings. This makes LIF units more sensitive when used to characterize soils contaminated with diesel fuel or heating oils. The systems have reduced sensitivities when using lighter fuels such as gasoline, JP-4 and JP-5, which contain smaller quantities of multiringed PAHs.

2.2 Characterization of Site Geology and Hydrogeology

The geologic and hydrogeologic characteristics affect the migration of contaminants during extraction and under passive-migration conditions. Proper characterization of the geology and hydrogeology at the site is necessary because the information may be used to predict the success of LNAPL-recovery efforts at the site. Sections 2.2.1 and 2.2.2 describe the required data to accurately characterize the geology and hydrogeology at a potential site.

2.2.1 Site Geology Analyses

The soil types and site geologic conditions are controlling factors in the potential migration and recovery of LNAPL. Soil types and site geologic conditions, such as stratigraphy, porosity, and permeability, define the physical framework in which the bioslurper will operate. The physical framework refers to the distribution and configuration of the aquifer of interest and any related stratigraphic units, such as confining layers, if present. Of particular interest are the thickness, continuity, lithology, and geologic structure of units that are relevant to the operation of the bioslurper. The physical framework of a flow system can be defined using the concept of hydrostratigraphic units, which consist of geologic units, or formations, of similar hydrogeologic properties. Several geologic formations may be combined into a single hydrostratigraphic unit, or a geologic formation may be subdivided into aquifers and confining units.

Definition of the physical framework and hydrostratigraphic units can be done using traditional geologic techniques. Geologic maps and cross sections showing the areal and vertical extent and boundaries of the formations are necessary to identify the hydrostratigraphic units of interest and define the physical extent of the system. Topographic maps showing surface water bodies and potential hydrologic divides provide information regarding potential LNAPL/surface water/groundwater interactions. Contour maps depicting the elevation of the base of the aquifers and confining beds, and isopach maps showing the thicknesses of the aquifers and confining beds, also aid in defining the physical extent of the units.

Geologic and topographic maps of the site of interest typically are available through state or federal agencies, including the state geological survey and the U.S. Geological Survey. Additional references may be available regarding the geology or water resources of the area. These information sources, if not already obtained, should be reviewed for pertinent information. Site-specific information, such as lithologic logs, geophysical logs, and soil boring descriptions, should be used to develop site-specific geologic cross sections, contour maps depicting the physical boundaries of the system, and isopach maps defining the thicknesses of the units of interest. If this information is not available, it is recommended that several boreholes be drilled and sediment/rock cores be collected and logged to adequately characterize the stratigraphy at the site.

Soil samples representative of each important lithology present at the site should be collected for determination of porosity and grain-size distribution. Soil samples also may be used to develop soil-moisture retention curves, measure organic-carbon content, and perform batch sorption experiments to determine the amount of sorption expected at the site.

Geologic conditions favorable to hydrocarbon recovery through bioslurping are found at sites with geologic materials that are fairly homogeneous, with high porosity and relatively high permeability, such as sands and gravels. However, extremely coarse geologic materials may cause the vacuum to short-circuit during bioslurping if the water table at the site is shallow. Fine-grained confining layers in the unsaturated zone may increase the area of influence from the extraction well by containing the vacuum.

2.2.2 Site Hydrogeology

Proper characterization of the hydrogeologic framework of a site is equally important to the operation of a bioslurping system. Definition of the hydrogeologic framework is dependent on the description of the physical framework. The physical framework, or hydrostratigraphy, is used to form the physical structure of the conceptual model of the site, whereas the hydrogeologic information is used to conceptualize the movement of LNAPL and groundwater through the system. Information and interpretations needed to describe the movement of groundwater and hydrocarbons through the system include:

- ❑ Measurement of water-level elevations and the construction of hydrographs
- ❑ Measurement of air/oil and oil/water interface elevations (Section 2.1.2.1)
- ❑ Estimation of LNAPL thickness and total volume of hydrocarbons (Section 2.1.2.5)

- Development of potentiometric-surface or water-table maps
- An assessment of the direction of groundwater flow
- Determination of the hydraulic gradients across the site
- Measurement of the hydraulic conductivities of the aquifer materials
- Calculation of the average velocity of groundwater flow
- Identification and quantification of recharge, discharge, and flow boundaries.

Water-level measurements from wells and piezometers within the area of interest can be used to determine the direction of groundwater flow, the location of recharge and discharge points, the connection between individual aquifers, and the hydraulic communication between aquifers and surface-water bodies. It is essential to have as much water-level data from as many wells and piezometers at the site as possible, because water-table elevations and, consequently, hydraulic gradients and groundwater flow directions, can change considerably over a short distance in the aquifer. It is also important to evaluate water-level data from several time periods throughout the year because water-table elevations often fluctuate temporally due to changes in precipitation, groundwater recharge, surface-water elevations, and pumping stresses. At a minimum, quarterly water-level measurements over a period of one year should be evaluated. Hydrographs, or plots of water-level elevation versus time, are useful in assessing the amount of variability in the water-table elevation in a particular well over time. It is essential to have accurately surveyed well and piezometer locations (easting and northing coordinates), as well as reference elevations from the tops of the well casings. The well coordinates and reference elevations are necessary for determining the water-table, air/oil, and oil/water elevations and for estimating the total LNAPL volume.

Air/oil and oil/water interface measurements, or interface contact measurements from wells and piezometers within the area of interest, can be used to determine the extent of LNAPL contamination and estimate the total volume of LNAPL present at the site. Estimation of LNAPL volume is discussed in Section 2.1.2.5. As with water-level data, it is essential to have as much air/oil and oil/water interface data from as many wells and piezometers at the site as possible. LNAPL elevations and thicknesses tend to vary both spatially and temporally. In addition, a fluctuating water table may influence LNAPL thickness by causing a smear zone in which LNAPL is distributed.

The hydraulic gradient is the change in hydraulic head (water-table elevation) divided by the length of groundwater flow along a flowpath. Accurate assessment of the hydraulic gradients across a site is dependent on the measurement of water levels in as many monitoring wells and piezometers as possible. Because hydraulic gradients can vary spatially and fluctuate temporally, it is essential to have as much water-level data as possible. Sites near surface water bodies such as lakes or rivers are more likely to be affected by seasonal variations in water-table elevations, changes in hydraulic gradients, and changes in groundwater flow directions. In addition, sites near oceans or harbors may be influenced by tidal patterns that impact water-table elevations and hydraulic gradients.

Hydraulic conductivity is a measure of an aquifer's ability to transmit water and is expressed as the rate at which water can move through the porous medium. Hydraulic conductivity is perhaps the most important parameter governing LNAPL and groundwater flow. Conceptually, the hydraulic conductivity of an aquifer is the volumetric flowrate that the aquifer will permit through a unit surface area under a specified hydraulic gradient. The units used to describe hydraulic conductivity are derived from units of volumetric flow normalized to surface area, or (length^3/t)/length^2, which reduces to length/t. Common units for hydraulic conductivity are cm/s and ft/d.

Darcy's law describes the relationship between hydraulic conductivity and the volumetric flowrate, hydraulic gradient, and surface area associated with the flow of water through a porous medium, and can be expressed as:

$$Q = K\frac{dh}{dl}A \tag{2-2}$$

where: Q = volumetric flowrate (length^3/t)
K = hydraulic conductivity (length/t)
dh/dl = hydraulic gradient
A = cross-sectional area of flow (length^2).

Porosity is defined as the ratio of the void spaces in a unit of soil or rock to the total volume of that unit. Porosity is usually expressed either as a percent or in decimal fraction (Freeze and Cherry, 1979). Table 2-2 lists representative ranges of porosity for different lithologies.

The ability of a rock or sediment to transmit LNAPL and water is dependent on the porosity, either primary or secondary, and the permeability or interconnectedness of the pore spaces. Some consolidated, but poorly cemented, sedimentary rocks, such as coarse sandstone and some limestones, will readily transmit water through the pore spaces surrounding individual grains of the rock matrix. Well-cemented, or indurated, sedimentary

TABLE 2-2. Ranges of porosity for various soil and rock types (Freeze and Cherry, © 1979. Reprinted by permission of Pearson Education, Inc., Upper Saddle River, NJ 07458)

Soil or Rock Type	Porosity (%)
Unconsolidated Deposits	
Gravel	25–40
Sand	25–50
Silt	35–50
Clay	40–70
Rocks	
Fractured basalt	5–50
Karst limestone	5–50
Sandstone	5–30
Limestone, dolomite	0–20
Shale	0–10
Fractured crystalline rock	0–10
Dense crystalline rock	0–5

formations (e.g., shale and crystalline limestone) and metamorphic and igneous formations all typically have low primary porosities, and depend more on secondary porosity resulting from fractures, joints, and chemical solution to transmit groundwater. Shales and some clays have relatively high primary porosities, but the clay minerals are platelike and overlap to the degree that the very small pore spaces are not interconnected. Shales and some clays must therefore be fractured or jointed to function as an aquifer. Unconsolidated formations are characterized by a lack of cementation binding individual grains of the matrix. Unconsolidated formations range from well-sorted recent alluvial and outwash deposits to poorly sorted glacial till. Well-sorted deposits have average primary porosities, and the pore spaces are more interconnected than in poorly sorted materials.

The effective porosity, commonly referred to in hydrogeology, is the volume of the void spaces through which water or other fluids can travel in a rock or sediment divided by the total volume of the rock or sediment, or the porosity available for fluid flow. The effective and total porosities can be determined in the laboratory using samples collected during drilling.

2.2.3 Fluid Characterization

In addition to the geologic and hydrogeologic data that should be collected at a site being considered for bioslurping activity, information should be obtained about the character of the LNAPL present in the flow system.

Along with air/oil and oil/water interface measurements, samples of the LNAPL should be obtained and analyzed for density and viscosity. If possible, surface tension data should be obtained for use in (1) determining the relevant scaling parameters used to estimate the total LNAPL volume and (2) computer modeling of LNAPL migration and recovery. Samples of the LNAPL present at the site also may be used in laboratory batch or column experiments to evaluate the behavior of the oil as it contacts soil at the site. Estimates of sorption and residual oil saturation can be made using soil samples and LNAPL samples collected from the site. LNAPL samples also should undergo chemical analyses to determine the general composition of the LNAPL. If data cannot be measured through laboratory measurements, Table 2-3 may be used to estimate the chemical and physical characteristics of the LNAPL. Table 2-3 presents the characteristics of unweathered fuels.

TABLE 2-3. Characteristics of unweathered fuels

	JP-5	Gasoline	JP-4	Diesel
Density (g/mL)	0.819[(a)]	0.7-0.8[(b)]	0.751-0.802[(c)]	0.810-0.936[(c)]
Viscosity (N-s/m^2)	1.49E-03[(a)]	5.29E-04[(a)]	N/A	N/A
Surface tension (N/m)	2.68E-02[(a)]	2.26E-02[(a)]	2.50E-02[(a)]	N/A
Boiling-point distribution				
<C8	2.0	69.0	62.7	0.5
C9	8.1	12.4	15.5	3.6
C10	20.7	10.4	10.2	8.6
C11	25.2	5.2	4.2	9.6
C12	21.0	2.3	2.9	10.9
C13	14.0	0.7	2.3	10.9
C14	6.7	N/A	1.4	11.3
C15	2.3	N/A	0.8	9.7
>C16	N/A	N/A	N/A	34.9
Log K_{ow}	3.3-7.06	2.13-4.87[(b)]	3-4.5[(c)]	3.3-7.06[(d)]
Log K_{oc}	3.0-6.7	1.81-4.56[(b)]	N/A	3.0-6.7[(d)]
Henry's law constant (atm-m^3/mol)	5.9E-5–7.4	4.8E-4–3.3[(b)]	1E-4–1E+1[(c)]	5.9E-5–7.4[(d)]
Water solubility (mg/L)	5	insoluble[(b)]	57[(c)]	5[(d)]
Vapor pressure (atm)	4.61E-02[(a)]	N/A	0.12[(c)]	2.79E-3–3.47E-2[(d)]

N/A = not available.

(a) Vargaftik, N.B. 1975. *Handbook of Physical Properties of Liquids and Gases—Pure Substances and Mixtures.* 2nd ed. Hemisphere Publishing Corporation, New York, NY.

(b) U.S. Department of Health and Human Services. 1995. *Toxicological Profile for Automotive Gasoline.* U.S. Department of Commerce, National Technical Information Service, Washington, DC.

(c) U.S. Department of Health and Human Services. 1995. *Toxicological Profile for Jet Fuels (JP4 and JP7).* U.S. Department of Commerce, National Technical Information Service, Washington, DC.

(d) Agency for Toxic Substances and Disease Registry. 1993. *Toxicological Profile for Fuel Oils.* U.S. Department of Commerce, National Technical Information Service, Atlanta, GA.

2.3 Evaluation of Site Characterization Data

Site characterization is an important component for preparing bioslurper operations at the site. The site characterization data may be used to evaluate the proper location for the LNAPL recovery pilot test, the potential for LNAPL recovery at the site, preliminary data for full-scale design, the most appropriate time of the year to conduct the pilot test, and a baseline evaluation of the contamination that may be used for site closure. The data also may suggest that LNAPL-recovery testing is not recommended.

In general, the extraction well for the LNAPL recovery pilot test should be located in an area of the plume with the greatest LNAPL thickness (corrected as described in Section 2.1.2.1) and near or in wells displaying the most rapid LNAPL recovery rates during the baildown tests. If there is large variability in the thickness of the LNAPL layer, pilot testing at several wells is recommended. The concentration of contaminants in the vadose zone also should be taken into consideration when determining the location of the pilot test. If possible, the pilot test well and monitoring points should be located in soils that have relatively low O_2 (<5%) and relatively high TPH soil-gas concentrations. By properly locating the extraction well and soil-gas points, more accurate data regarding LNAPL recovery rates and vadose zone biodegradation rates may be generated.

The data from the fluid-level measurements and baildown testing provide an indication of the lateral extent of the LNAPL plume, and help to define LNAPL recoverability. Data from the baildown test can be used to evaluate the potential for LNAPL recovery at a specific well. If the well will not produce at least two borehole volumes of LNAPL within a 12-hour period, LNAPL recovery may not be economically feasible from that well. Baildown test data from prospective bioslurper sites indicate that wells displaying LNAPL-recovery rates of <0.1 gallon per day (gpd) generally do not perform well as LNAPL recovery wells. Therefore, other non-LNAPL recovery technologies, such as bioventing, should be investigated for site remediation if all the wells at the site produce less than 0.1 gpd of LNAPL during each baildown test. As noted in Section 2.1.2.2, temporal variations in the passive migration of LNAPL may occur. Therefore, it is important to determine if this variation occurs before eliminating the use of any technology.

Site characterization activities provide data that are useful for full-scale design. The LNAPL plume delineation may be used, along with the bioslurper ROI testing, in well locations and construction details for full-scale implementation of the bioslurping technology. Also, accurate characterization of contaminant variability and site conditions may help in evaluating the results of the pilot testing. For example, if a pilot test is performed at one well and that well happens to be in a highly contaminated area, the TPH concentrations in the stack gas and the aqueous discharge are likely to be high. However, if site characterization data suggest that contaminant levels at the remainder of the site are much lower, then the purchase of expensive treatment equipment for the off-gas and the aqueous discharge may be avoided. As mentioned previously, if there is large variability in the LNAPL thickness or in the geologic/hydrogeologic conditions at the site, multiple pilot tests are recommended for the site.

Historic groundwater and LNAPL data may be used to determine the most appropriate time of the year to conduct a pilot test. For example, if LNAPL was observed in the monitoring wells when the water table is low or if the passive migration of LNAPL during baildown tests is more rapid during the dry months, it would be most appropriate to conduct the pilot testing during those months. If the test can be performed only during the season when the water table is high, LNAPL may be trapped beneath the water table and affect the placement of the drop tube in the extraction well. By lowering the water table, the previously trapped LNAPL may become mobilized and captured during the pilot test. Additionally, the historic site characterization data for the LNAPL plume and groundwater migration may provide valuable information regarding why a pilot test was unsuccessful.

LNAPL volume estimates, apparent LNAPL thickness measurements, plume delineation, and vadose zone contaminant analyses performed during the site characterization provide a baseline that may be used to monitor the progress of the LNAPL removal project. Such data also may be used when the site is being negotiated for closure. If reliable LNAPL volume estimates were determined, the duration of the LNAPL recovery project may be estimated using the LNAPL recovery rates observed during the pilot test and full-scale operation. As stated previously, periodic in situ respiration testing may eliminate the need for frequent soil sampling and analysis. Because the respiration tests may be used to monitor the progress of the remediation project, soil samples need to be collected only at the beginning and end of the project.

2.4 References

DAEM, see Draper Aden Environmental Modeling, Inc.

de Pastrovich, T.L., Y. Baradat, R. Barthel, A. Chiarelli, and D.R. Fussell. 1979. *Protection of Groundwater from Oil Pollution.* CONCAWE (Conservation of Clean Air and Water-Europe). The Hague, Netherlands. 61 pp.

Draper Aden Environmental Modeling, Inc. 1997. *OILVOL: Oil Volume Estimator. Technical Documentation and User's Guide*. Blacksburg, VA. 11 p.

Farr, A.M., R.J. Houghtalen, and D.B. McWhorter. 1990. "Volume Estimation of Light Nonaqueous Phase Liquids in Porous Media." *Groundwater, 28*(1): 48-56.

Freeze, R.A., and J.A. Cherry. 1979. *Groundwater*. Published by Prentice-Hall, an imprint of Pearson Education, Inc., Upper Saddle River, NJ. 604 p.

Leeson, A., and R.E. Hinchee. 1997. *Soil Bioventing Principles and Practices*. CRC Press Lewis Publishers, Boca Raton, FL. pp. 244.

Lenhard, R.J., and J.C. Parker. 1990. "Estimation of Free Hydrocarbon Volume from Fluid Levels in Monitoring Wells." *Groundwater, 28*(1):57-67.

Mercer, J.W., and R.M. Cohen. 1990. "A Review of Immiscible Fluids in the Subsurface: Properties, Models, Characterization and Remediation." *Journal of Contaminant Hydrology, 6*: 107-163.

3
Pilot Testing

This chapter describes the methods used to conduct a pilot-scale test for LNAPL recovery, including test wells and equipment that are required to conduct field treatability tests. An LNAPL-recovery pilot test is performed to compare the cost-effectiveness of bioslurping with that of conventional LNAPL-recovery technologies (i.e., skimming and drawdown). Bioslurping has been demonstrated to be the most effective technology at recovering free product. However, the bioslurper will recover groundwater while extracting the LNAPL and may not be cost-effective to operate at sites where water treatment is very expensive. Also, the pilot test may provide estimates of long-term LNAPL recovery rates. Using the estimated LNAPL recovery rates, bioslurping may prove to be very cost-effective due to the reduced operation time to remove the recoverable LNAPL, thus reducing the O&M costs.

Site-specific flexibility will be required and details will vary. To the extent possible, Sections 3.2, 3.3, and 3.5 identify equipment or system components that may be used under a variety of site conditions. The LNAPL recovery pilot test may be tailored to fit the objectives for site remediation or closure. For example, the bioslurper may be used to evaluate the feasibility of recovering LNAPL at the site. At a site in southern California, the bioslurper system was operated strictly in a bioslurping configuration to demonstrate that the LNAPL was not recoverable. Although a small amount of LNAPL was present in the monitoring wells (before the baildown test), no LNAPL was recovered during the three-week pilot test. The LNAPL was then believed to be unrecoverable because the bioslurper technology is the most effective method of LNAPL removal. Some of the information presented in this chapter was obtained from the *Test Plan and Technical Protocol for Bioslurping* (Leeson et al., 1995).

3.1 Permitting for Bioslurper Pilot Testing

Regulatory approval may be required for any or all of the following items prior to conducting a bioslurper pilot test: vapor discharge, aqueous discharge, system construction, and well installation. However, permitting requirements may not be required or may be waived for short-term pilot testing that generates relatively low contaminant releases.

3.1.1 Vapor Discharge Permits

In general, vapor discharge permits are waived for bioslurper pilot testing due to the low mass discharge rates of TPH into the atmosphere. If permits are required they are usually based on mass discharge limits for TPH and benzene. Most states limit the mass of TPH released per day or year. Typical discharge limits are approximately 5 lb/day. A limited number of states base their regulatory requirements on health risk. When a health risk approach is used, the regulatory agency limits the maximum downwind TPH concentration. For these regulatory agencies it may be necessary to perform an air dispersion model such as SCREEN3. If the regulatory agency will not waive the permitting requirements, treatment of the vapor discharge stream with activated carbon is likely to be the most cost-effective approach for pilot testing.

3.1.2 Aqueous Discharge Permits

An aqueous discharge permit may be required for the release of the bioslurper process water at some locations. Due to the low groundwater recovery rates and the relatively low contaminant concentration in the process water, the process water may be approved for release to the local wastewater treatment plant (WWTP) at most locations. When discharge to the WWTP is acceptable, the pilot-scale system should be properly located so access can be gained to a WWTP manhole or pipeline. At remote locations, it may be necessary to discharge the process water to the surface drainage system. In these cases, a National Pollution Discharge Elimination System (NPDES) permit will be required, and the discharge water must meet the requirements of the permit. If treatment is required, a clay/anthracite mixture can be used to remove the suspended petroleum hydrocarbons and the water can then be polished with activated carbon. This type of treatment should only be used for short-term pilot testing. The operation and maintenance of the clay/carbon treatment systems can be very time-intensive for long-term operation.

3.1.3 Well Installation Permits

If new fuel recovery wells are installed for the bioslurper pilot test, it may be necessary to apply for a permit for the installation of such wells. Typically, well installation is regulated at the state level; however, some states have water conservation districts that regulate well installation. The states or water conservation districts may also limit the volume of water extracted during the pilot testing.

3.1.4 Other Permits

Each state and each facility have specific guidelines and regulations for the construction, installation, and operation of environmental remediation systems. For example, some facilities may require some form of construction permitting for any system that will be at the facility for longer than two weeks. Most facilities require digging/drilling permits prior to any excavation or drilling operations. Other site-specific permits may be required (e.g., groundwater recovery permits), so it is necessary to inquire about all permits with the contact person at the facility.

3.2 Bioslurper Extraction Wells

In general, extraction wells for the pilot test should be located within the area of the LNAPL plume having greatest free-product thickness and the highest TPH concentrations in the vadose zone. If there is a large degree of variability in the geology or in the extent of contamination at the site, it may be necessary to perform additional pilot tests at different locations.

Construction of the bioslurper extraction wells may have significant effect on the LNAPL recovery rate and the recovery ratios of LNAPL/water and water/soil gas. The bioslurper system generates a vacuum that is evenly distributed inside a well casing from the top of the well to the air/LNAPL interface. This vacuum is released to the formation throughout the entire length of the screened section. Most of the vacuum is lost to the vadose zone, but some of the vacuum is used to pull LNAPL and groundwater into the well. Therefore, the optimum well design should minimize the amount of energy lost to the vadose zone and groundwater pulled into the well while maximizing the amount of LNAPL moved toward the wells.

In most cases, the extraction wells should possess a minimal length of screen both above and below the water table. However, the screened section should be long enough to cover the mobile LNAPL layer in the formation, which may fluctuate when the water table fluctuates. This design not only maximizes the available vacuum on the LNAPL layer, thus increasing the ROI with respect to the free product, but also reduces the amount of water and soil gas extracted by limiting the recharge area of the well. Results of tests conducted at bioslurper sites suggest that an extraction well with 3 ft of screen both above and below the water table is sufficient.

Factors that must be taken into account during the design of the extraction wells include the depth to the water table, seasonal fluctuations, tidal fluctuations, permeability of the formation (water and soil gas), and free-product thickness. At sites with a shallow water table (i.e., <7 ft below ground surface [bgs]), a shorter length of screen above the water table is preferred. If possible, the top of the screened section should be at least 4 ft bgs. Decreasing the length of screen above the water table at sites with a shallow water table greatly reduces the chance of short-circuiting atmospheric air to the extraction well. At sites where the water table is deep, it may be necessary to increase the length of screen above the water table to allow enough soil gas to be pulled into the system to provide air-entrainment of the fluids in the drop tube. Data from a soil-gas permeability test may be used to calculate the maximum soil-gas flowrate to the well. If the water table is deep (~50 ft bgs) and the formation is composed of medium-sized sand, the length of screen above the water table should be increased to 5 ft.

Seasonal and tidal water table fluctuations should be considered when designing extraction wells. In general, the extraction wells should be designed with the bottom of the screened section located at least 1 ft below the lowest groundwater level and the top of the screened section at least 1 ft above the highest water level. During bioslurper operations, the groundwater level in the well generally is maintained at the bottom of the drop tube, even if the groundwater level in the formation rises significantly. Similarly, the vacuum in the well tends to lift the groundwater level in the well even if the groundwater level in the formation drops below the bottom of the drop tube. However, lifting of the groundwater table would create a mound around the extraction well, thereby possibly driving the LNAPL away from the extraction well.

In some cases, it may be possible to use existing monitoring wells for the LNAPL recovery pilot test. If no suitable monitoring well exists, it will be necessary to install extraction wells for the test. The specifications for constructing an LNAPL-extraction well are described as follows:

- The recommended diameter of a well casing is between 2 and 6 inches, depending on material cost and equipment availability. Increasing the diameter of the well casing beyond 6 inches may increase the groundwater recovery rate significantly but may increase the LNAPL recovery rate only slightly.

- The bioslurper well casing should be constructed of schedule 40 polyvinyl chloride (PVC) and screened with a slot size that will allow soil gas to flow into the well while minimizing transport of fine soil particles into the well. The slot sizes generally range from 0.006 to 0.020 in (#6 to #20 slots). The length of the screened interval is discussed earlier in this chapter.

- The recommended drilling method is hollow-stem auguring. Whenever possible, the diameter of the annular space should be at least two times greater than the extraction well outside diameter. The annular space corresponding to the screened interval should be filled with silica sand or equivalent. The annular space above the screened interval should be sealed with wet bentonite or grout to prevent short-circuiting of air from the surface. Figure 3-1 shows a typical bioslurper well.

As discussed in Section 2.1.2.1, LNAPL tends to accumulate in an extraction well and depresses the groundwater level in the well. This fact should be taken into account when designing the extraction wells. For most contaminated sites this effect can be ignored, but at sites with thick free-product layers, this issue should be considered during construction of the extraction wells.

3.3 Soil-Gas Monitoring Points

Soil-gas monitoring points are used for pressure measurements and soil-gas sampling. They generally are installed at three or more depths and at least three locations. The monitoring points should be located in contaminated soil with >1,000 mg/kg of TPH. However, it may not be possible to locate all monitoring points in contaminated soil, especially the points farthest from a bioslurper well. In this case, it is important to ensure that the point closest to the extraction well is located in contaminated soil, and that the intermediate point, if possible, also is placed in contaminated soil. If no monitoring points are located in contaminated soil, meaningful in situ respiration test results cannot be derived. Based on Battelle's experience, to ensure successful in situ respiration testing, monitoring points must have significant TPH concentrations (ideally >10,000 ppmv) and low oxygen

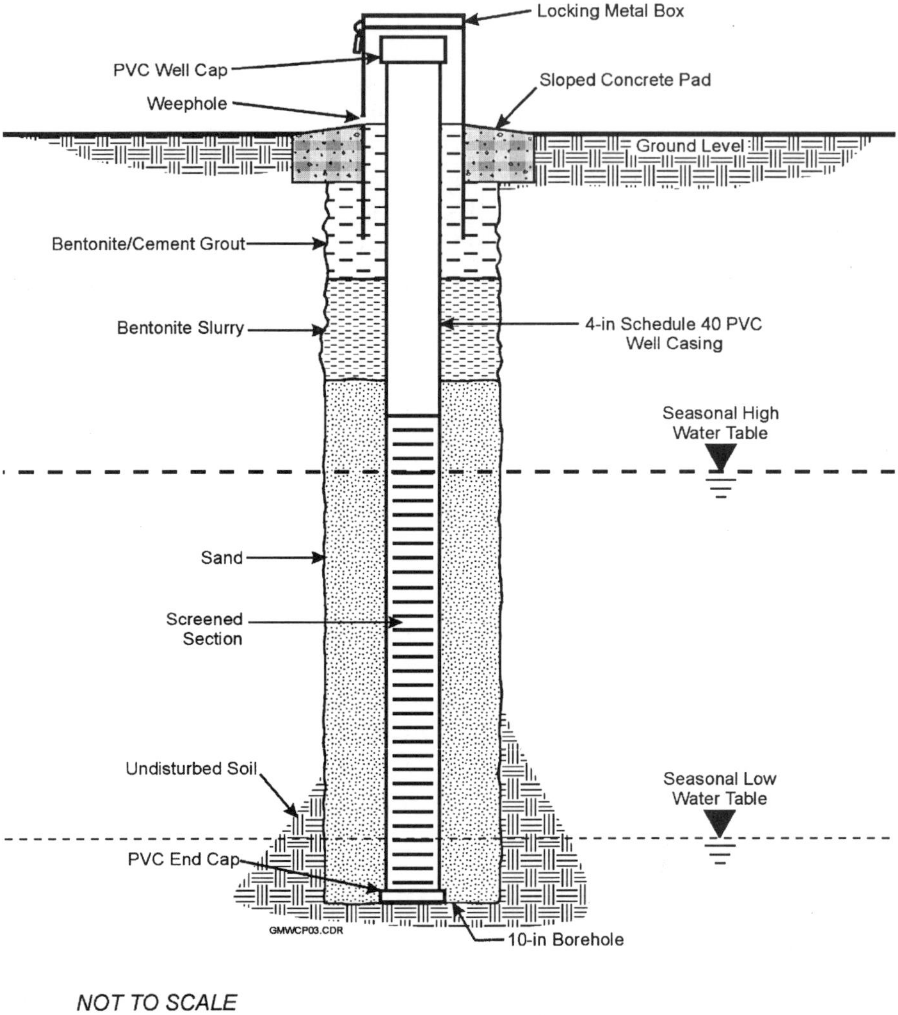

FIGURE 3-1. Diagram of a typical bioslurper well

concentrations (ideally 5% O_2 or less) in soil gas. A background soil-gas monitoring point also is needed to establish background soil-gas concentrations. This monitoring point may be an existing monitoring point or monitoring well in an uncontaminated location.

3.3.1 Locations of Monitoring Points

In general, the three monitoring points should be located in a straight line radially out from the extraction well. Figure 3-2 displays a conceptual

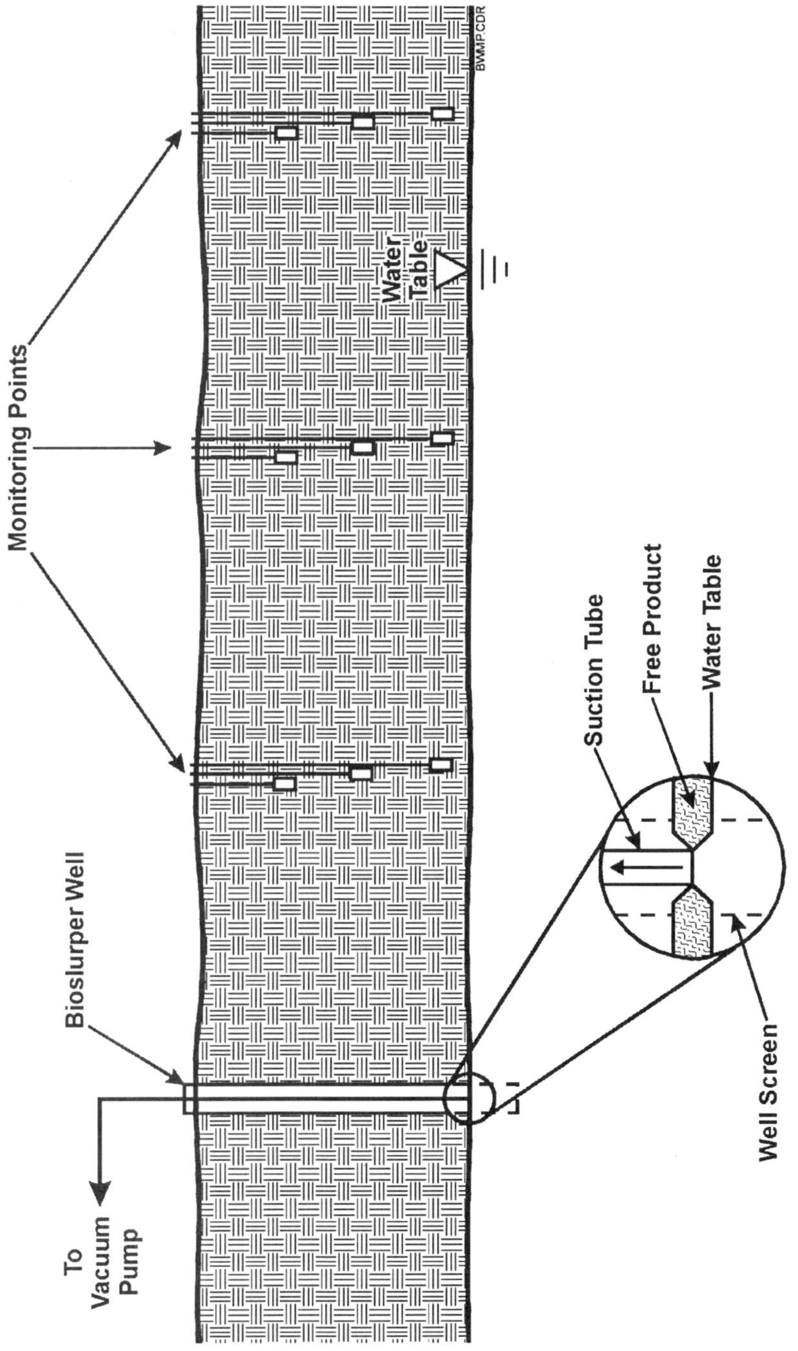

FIGURE 3-2. Conceptual configuration of soil-gas monitoring points and bioslurper well

configuration of the soil-gas monitoring points. Typically, the lateral spacing from the extraction well is determined based on the site geology and the depth to the top of the extraction well screen. The lateral spacing generally decreases with increasing soil grain size (or increasing soil permeability), but increases with increasing depth to the top of the well screen or if low-permeability horizons exist in the vadose zone above the contamination. Additional monitoring point locations may be needed for a variety of site-specific reasons including, but not limited to, spatial heterogeneity, obstructions (buildings, underground tanks, etc.), and a desire to monitor a specific location.

Each monitoring point generally is screened to at least three depths. The deepest screen should be placed approximately 1 ft above the water table or liquid interface. Consideration also should be given to potential seasonal water table fluctuations and soil types. In more-permeable soils, the monitoring points can be screened closer to the water table. In less-permeable soils, they should be screened farther above the water table. The shallowest screen usually is placed 3 to 5 ft bgs. The intermediate screen is positioned at a depth with an equal distance between the deepest and the shallowest screens. It is generally a good practice to place the intermediate screen within the upper portion of the screened section of the bioslurper extraction well to maximize its pressure-monitoring capabilities.

Using these rules of thumb, in a sandy soil with a groundwater depth of 15 ft and a bioslurper well screened from 10 to 20 ft bgs, acceptable screen depths for a soil-gas monitoring point would be 14, 10, and 5 ft. It may be necessary in some cases to install additional screen depths to ensure that contaminated soil is encountered, to monitor differing stratigraphic intervals, or to adequately monitor deeper sites with broadly screened bioslurper wells. In all cases, consideration should be given to place monitoring points in distinct lithologic units. Screened intervals should be placed in more-permeable soil layers, and low-permeability zones should be avoided if possible.

3.3.2 Monitoring Point Construction

Figure 3-3 presents a typical soil-gas monitoring point. A monitoring point generally consists of three small-diameter (e.g., ¼-inch) nylon tubes, each extending from the ground surface to a specified depth where a gravel-filled screen (1 inch diameter and approximately 6 inches long) is connected. The screen is surrounded with a sand pack, which normally extends for 3 inches both above and below the screened section. In low-permeability soils, a longer sand pack may be desirable. In wet soils, a longer sand pack with the screen near the top also may be desirable. A bentonite seal at

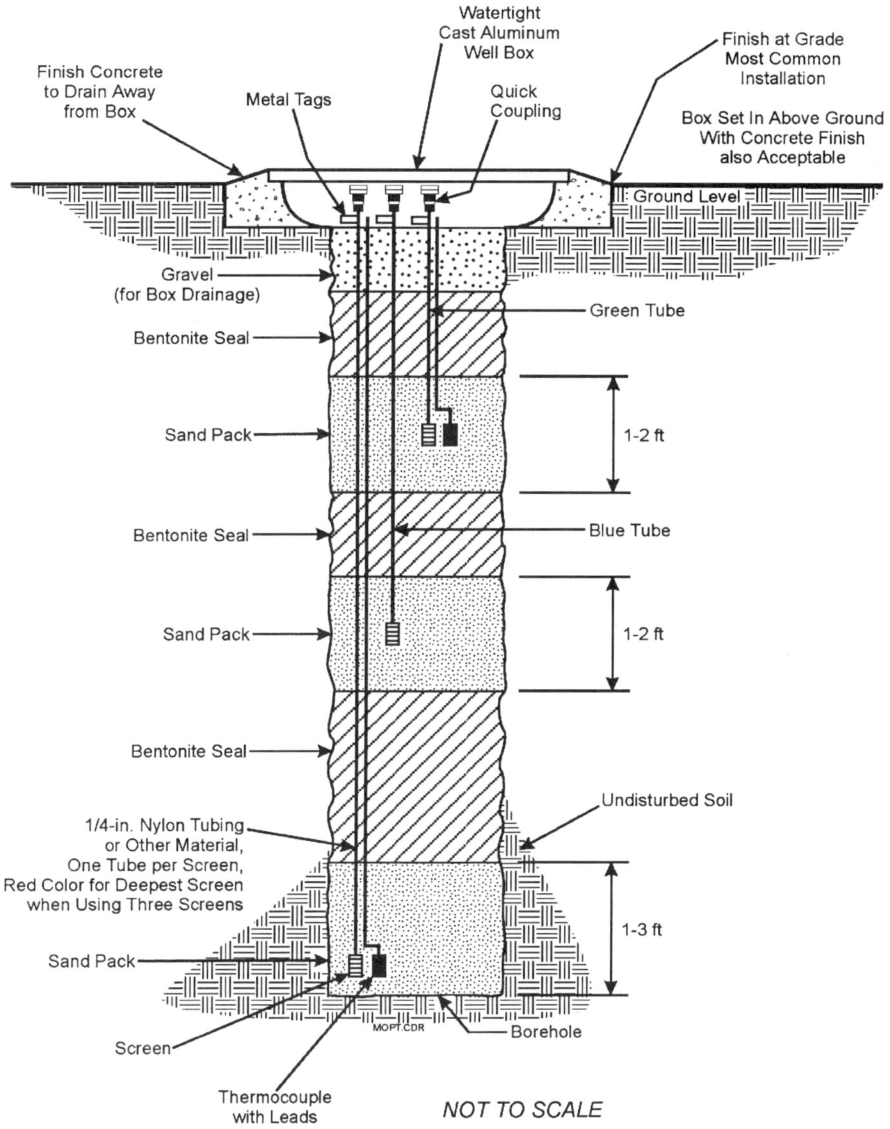

FIGURE 3-3. Diagram of a typical soil-gas monitoring point

least 2 ft thick is required both above and below each sand pack to ensure that pressure and soil-gas samples taken are discrete to that depth.

Tubing used for the monitoring points must have sufficient strength and be nonreactive. Nylon tubing has been used in most cases and does not appear to have any problems with surface interactions. All tubing from each screened section may be connected with quick-connect couplings, which are placed and protected in a watertight cast-aluminum well box on the ground surface. Each screened depth may be labeled with a code, such as MP (for monitoring point)–[depth to bottom of screened interval], on a metal tag placed close to the ground surface or in the watertight cast-aluminum well box.

In most cases, Type-K (or equivalent) thermocouples are installed side-by-side with the gravel-filled screens. The thermocouples are connected to a thermocouple thermometer. The thermocouple wires may be labeled using the same system used for the monitoring tubing, except that a two-letter abbreviation for thermocouple (TC) is added to the identification label.

3.4 Baseline Measurements

If deficiencies exist in the site characterization data, it will be necessary to perform on-site tests to adequately delineate the site conditions. The tests required to characterize a site are discussed in Sections 2.1 and 2.2 and are listed below:

- Interface measurements (Section 2.1.2.1)
- Baildown testing (Section 2.1.2.2)
- Soil-gas survey (Section 2.1.2.3)
- In situ respiration testing (Section 2.1.2.4)
- Soil/site geology analyses (Section 2.2.1)
- Site hydrogeology (Section 2.2.2).

The pertinent sections in Chapter 2 include brief protocols for the specific characterization tests.

3.5 Mobilization, Installation, and System Shakedown

The pilot-scale bioslurper system is mounted on a trailer that can be operated at multiple test locations. In most cases, the system can be shipped via road freight to the general test location. Once the system arrives near the site, it can be moved to the specific test location using a truck with a trailer hitch. After the trailer has been properly located, the electrical source should be connected to the LRP and any other equipment requiring power. Prior to initiating the LNAPL recovery test, a system shakedown should be performed to ensure that the system and the safety switches are functioning properly.

3.5.1 Pilot-Scale Bioslurper System

A pilot-scale bioslurper system is a trailer-mounted unit consisting of extraction pipe and hose, an equalization vessel, an LRP, an oil/water separator (OWS), and a surge tank with an automatic overflow shutoff switch. The system may include stack-gas and/or water treatment equipment, depending on requirements specified by the site and/or regulatory authorities. Figure 3-4 presents the schematic diagram of a typical pilot-scale system.

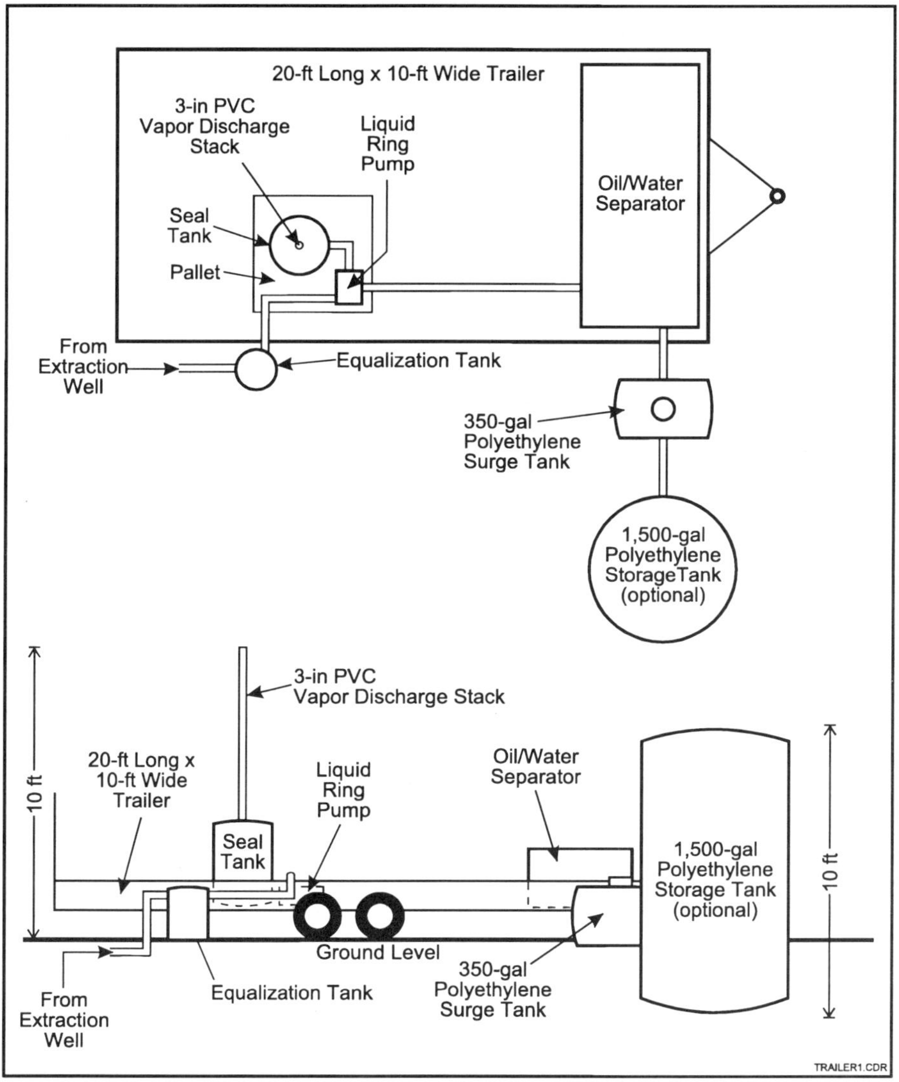

FIGURE 3-4. Trailer-mounted, pilot-scale bioslurper unit

In most cases, a 7.5-horsepower (hp) LRP is used to extract LNAPL, groundwater, and soil gas. A pump larger than 7.5 hp is not recommended for single-well pilot testing (except at sites with a water table more than 50 ft deep). If a larger pump is used, the soil-gas recovery rate may be relatively high and the size of the LRP required for full-scale bioslurper implementation may be overestimated (see Section 3.6.2.4 for proper quantification of stack-gas volume). LNAPL is separated from groundwater by passing the oil/water mixture through a conventional gravity OWS. If stable oil/water emulsions and floating solids (consisting of air, soil particles, and LNAPL) are formed (perhaps, through the slurping action in extraction wells and/or the mixing action in the LRP), an OWS with coalescer packings and a cone-shape bottom is recommended for separation. LNAPL, along with the floating solids, overflow to the oil reservoir prior to being transferred to a double-walled fuel-rated storage tank. The water from the OWS is stored in a storage tank, disposed of to a sewer, commercially disposed of, or further treated to remove stable oil/water emulsions.

The stack gas may be discharged directly to the atmosphere or treated with gas-phase activated carbon canisters plumbed in series to the bioslurper vapor discharge stack. A pressure gauge should be placed on the stack (if the stack gas is being treated with activated carbon) and vapor sampling ports may be placed before, between, and after the two carbon canisters. It may be necessary to install a regenerative blower between the LRP and canisters to provide enough pressure to push the stack gas through the carbon canisters. The life of the activated carbon may be extended if the stack gas from the bioslurper is passed through a dehumidifier prior to entering the activated carbon vessels. The discharge line from the second canister should be fitted with a pitot tube flow indicator.

3.5.2 Site Mobilization

As soon as the test location has been determined, the system must be mobilized and the site must be properly prepared to ensure safe and efficient operations. The activities to be completed may include, but not be limited to, the following items:

- ❑ Obtain required permits (such as digging permit, well installation permit, air/water discharge permit, etc.).
- ❑ Prepare a connection to electrical power (or use a generator).

- ❑ Construct secondary containment (if required).
- ❑ Install extraction wells/soil-gas monitoring points (if not yet in existence).
- ❑ Acquire a prefabricated, trailer-mounted bioslurper system, associated drums and water and fuel storage tanks, valves, piping, and miscellaneous materials and supplies.
- ❑ Obtain site support (if needed).

3.5.3 System Installation and Shakedown

The trailer-mounted bioslurper unit should be located at a level area near the extraction well(s). The system and the associated storage tanks may need to be placed in a contained area, depending on site-specific requirements. After the system is connected to the extraction well, a start-up test should be conducted to ensure that all system components are operating properly. Components to be checked include the extraction piping; LRP; aqueous effluent transfer pump; vapor, fuel, and water flowmeters; oil/water interface probes; soil-gas analysis instrumentation; emergency shutoff switches in the seal-water tank, OWS, liquid storage tanks, and annunciator panel; and any vapor/effluent treatment system components. The annunciator panel is used to indicate the cause of a system failure, such as a high-level situation in a liquid storage tank. If a shutdown occurs, a light corresponding to a specific emergency switch is extinguished. A checklist is provided in Figure 3-5 to document the system shakedown.

3.6 LNAPL Recovery Testing

After the system shakedown has been completed, testing can be performed to compare the feasibility of bioslurping with that of conventional free-product recovery techniques (i.e., skimming and dual-pump drawdown). Table 3-1 presents a recommended schedule for a comparison test of these technologies. It is important to follow the testing sequence in Table 3-1 by performing simulated skimmer testing before bioslurping and simulated dual-pump drawdown testing. The sequence will ensure minimum disturbance of the hydrogeologic conditions as the testing proceeds.

Checklist for System Shakedown

Site: ____________________

Date: ____________ Operator's Initials: _____

Equipment	Check if OK	Comments
Liquid ring pump		
Aqueous effluent transfer pump		
Oil/water separator		
Vapor flowmeter		
Fuel flowmeter		
Water flowmeter		
Emergency shutoff float switch effluent transfer tank		
Analytical field instrumentation O_2/CO_2 analyzer Hydrocarbon analyzer Oil/water interface probe Magnehelic boards Thermocouple thermometer		

FIGURE 3-5. Bioslurper pilot test shakedown checklist

TABLE 3-1. Schedule of activities

Pilot Test Activity	Test Duration
Mobilization	
Soil characterization (if not already performed)	
Product/groundwater interface monitoring	Depends on number of wells to be monitored
Baildown tests	1 day
Soil-gas survey (limited)	1 day
Simulated skimmer testing	2 days
Bioslurper testing	4 days (minimum)
Simulated drawdown testing	2 days
SVE testing (if indicated)	1 day

3.6.1 Simulated Skimmer Testing

A simulated skimmer test is performed using the bioslurper system to demonstrate LNAPL and groundwater recovery under passive recovery conditions. Figure 3-6 displays the wellhead configuration during the test. If available, a peristaltic pump or a conventional skimmer may be used to replace the LRP to extract the LNAPL. When using a LRP for testing, the well casing should be open to the atmosphere to ensure that no vacuum is generated in the well, and that LNAPL moves passively from the formation into the well. The drop tube should be placed at the oil/water interface in the well to prevent the formation of a cone of depression in the water table.

Throughout the test, LNAPL and groundwater recovery rates should be monitored. Vapor discharge rates and contaminant concentrations also should be monitored if a LRP is used to extract the fluids. If significant quantities of groundwater are removed from the well, the drop tube should be repositioned to prevent drawdown of the water level in the well.

3.6.2 Bioslurper Extraction Test

When the simulated skimmer test is complete, the ball valve at the wellhead should be closed to begin bioslurping (the location of the drop tube should remain at the oil-water interface) (see Figure 3-7). Meanwhile, initial soil-gas pressures should be taken at all soil-gas monitoring points. Sections 3.6.2.1 through 3.6.2.7 discuss the activities that must be completed during the bioslurping testing.

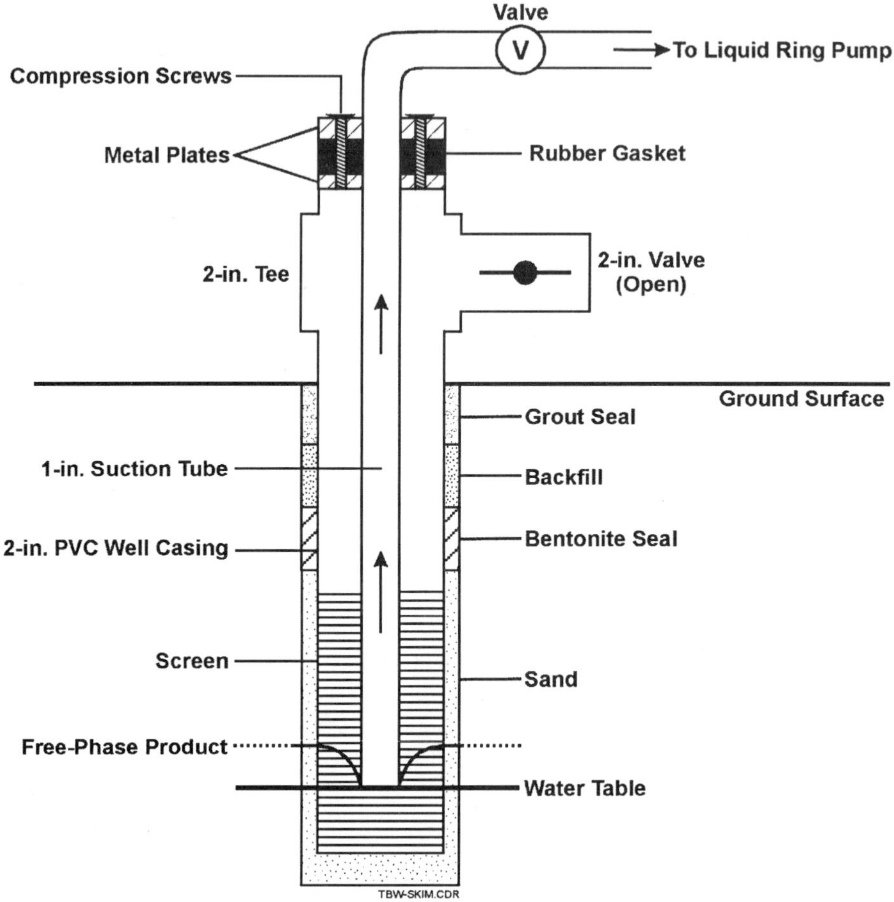

FIGURE 3-6. Configuration of drop tube and wellhead during simulated skimmer test

3.6.2.1 LNAPL Recovery Rate and Volume

The LNAPL recovered may need to be transferred from the oil reservoir on the OWS to a large holding tank using a hand-operated drum pump, or the LNAPL can be allowed to drain under gravity. The recovered volume may be quantified using an in-line flow-totalizer meter. It is recommended that the volume be measured every 30 minutes for the first 2 hours, every 2 hours for the next 10 hours, and every 12 hours until the test is complete. This procedure will make it easier to differentiate the initial slug of LNAPL recovery from sustainable LNAPL recovery.

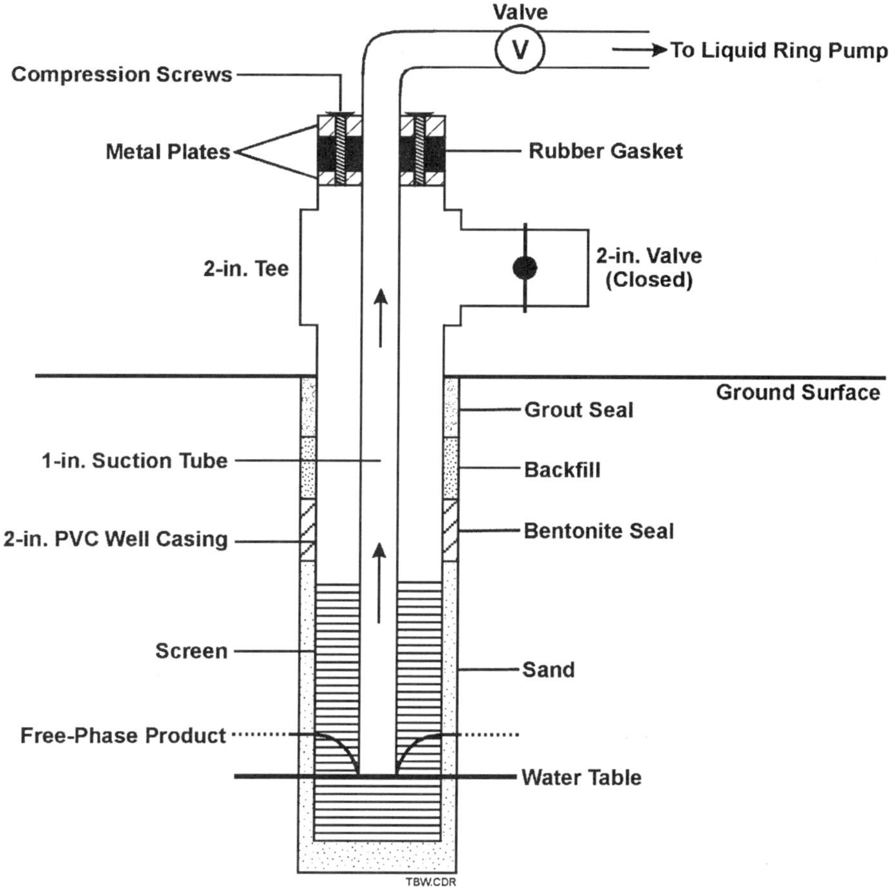

FIGURE 3-7. Configuration of drop tube and wellhead during bioslurper extraction test

3.6.2.2 Groundwater Recovery Volume

The groundwater recovery volume may be quantified using an in-line flow totalizer meter. The mass of petroleum hydrocarbons removed in the aqueous phase can be calculated based on the results of the effluent analysis (see Section 3.6.2.5) and the discharge volume.

3.6.2.3 Vadose Zone Radius of Influence

A vadose zone ROI test should be performed at the start of the bioslurper extraction test. The vacuum produced in the extraction well radiates into the vadose, capillary, and saturated zones of the formation. The vadose zone ROI is calculated using the magnitude of the vacuum in soil-gas monitoring points at various distances from the extraction well. The soil-gas pressure should be measured approximately every 2 hours until the change in pressure is less than 0.1 inch of H_2O. The pressure versus distance data are then plotted on an x-y plot, and a line is fitted through the data. The intersection of the line with the 0.1-inch H_2O vacuum is considered the vadose zone ROI.

3.6.2.4 Stack-Gas Volume

The flowrate of the stack gas may be quantified using a pitot tube flow indicator. The pitot tube is connected to a differential pressure gauge calibrated in inches of water. The flowrate in cubic feet per minute (cfm) is determined by referencing the differential pressure to a flow calibration curve (see Figure 3-8). The appropriate flow calibration curve can be obtained from the manufacturer of the pitot tube.

The volume of vapor discharge may be calculated based on the average flowrate and the hours of operation. The mass of petroleum hydrocarbons extracted in the vapor phase can be calculated based on the average concentration of two vapor samples taken (see Section 3.6.2.5) and the volume of soil gas extracted.

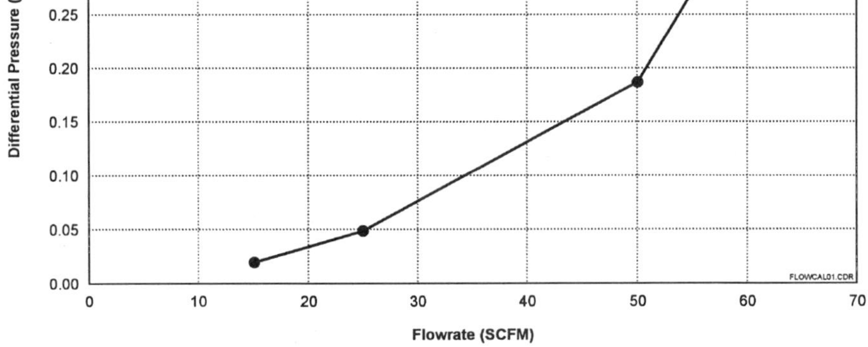

FIGURE 3-8. Typical flow calibration curve for the bioslurper vapor discharge pitot tube, 3-inch, schedule 80 PVC stack

If the stack-gas flowrate is relatively high (>15 cfm), the LRP may be extracting more soil gas than is required for liquid entrainment. In such cases, it is recommended that the soil-gas extraction rate be reduced (using the ball valve located at the top of the drop tube) to the minimum level required to lift fluids to the ground surface. The stack-gas flowrate should be remeasured at the restricted flowrate, and this flowrate may be used to estimate the minimum-sized LRP for full-scale operation. In addition, the system operating parameters (stack-gas TPH concentration, groundwater recovery rate, process water TPH concentration, and vadose zone ROI) should be remeasured or recalculated using the lower extraction rate.

3.6.2.5 Sampling and Analyses for Process Monitoring

Table 3-2 presents a typical sampling and analysis plan for monitoring the process performance during the pilot study. The plan involves taking a number of soil-gas, process water, and LNAPL samples for various measurements during the pilot test. The analytical methods, method detection limits (MDLs), suggested sample containers, sample sizes, preservation techniques, and holding times are listed in Table 3-2.

In general, at least two bioslurper stack-gas samples should be collected for benzene, toluene, ethylbenzene, and xylenes (BTEX) and TPH measurements. However, site-specific regulatory requirements may dictate the quantity and frequency of the stack-gas sample collection. The site regulators may require analyses for specific constituents. The samples are collected by connecting an evacuated 1-L Summa, polished stainless steel air-sampling canister to a sampling line on the stack. Prior to sampling, the sampling line should to be flushed with a representative vapor sample pulled from the stack by a vacuum pump. To collect a sample, the valve on the canister is opened, allowing the vacuum in the canister to be displaced with the vapor sample until atmospheric pressure is reached. The vacuum/pressure in each canister should be confirmed before and after each sampling event to ensure that the canister is received at the test site in an evacuated state and is completely filled during sampling.

Daily process water samples should be collected where the water exits the OWS. Additional water samples may need to be collected at the discharge point if the effluent from the OWS is further treated for stable oil/water emulsions. The water samples should be taken daily for BTEX, TPH (gasoline or diesel, depending upon the type of contaminants), and perhaps total iron and carbonate concentrations. Again, site-specific regulations may dictate the sample quantity and the constituents that must be analyzed. The samples should be observed to determine if floating solids

TABLE 3-2. Suggested sampling and analytical methods

Analysis	Analytical Method	MDL[a]	Container	Sample Size	Preservation Technique	Holding Time
			Soil			
Particle size distribution	ASTM[b] D422	NA[c]	Brass sleeve, polyethylene or glass container	1,000 g	Cool @ 4°C	180 days
Bulk density	ASTM D4531	NA	Brass sleeve, polyethylene or glass container	200 g	Cool @ 4°C	28 days
Porosity	ASTM D2434	NA	Brass sleeve, polyethylene or glass container	200 g	Cool @ 4°C	28 days
Moisture content	EPA 160.3	NA	Brass sleeve, polyethylene or glass container	50 to 300 g	Cool @ 4°C	28 days
BTEX	EPA 624/8240	10 μg/kg	Brass sleeve	100 g	Cool @ 4°C	14 days
TPH (jet fuel)	EPA Modified 8015/8240	10 mg/kg	Brass sleeve	100 g	Cool @ 4°C	14 days
TPH (diesel)	EPA 8015	10 mg/kg	Brass sleeve	100 g	Cool @ 4°C	14 days
			Soil Gas			
BTEX	EPA TO-3	1.0 ppbv[d]	Summa canister	1-L	NA	30 days
TPH (jet fuel)	EPA TO-3	10.0 ppbv	Summa canister	1-L	NA	30 days

TABLE 3-2. Suggested sampling and analytical methods (continued)

Analysis	Analytical Method	MDL[a]	Container	Sample Size	Preservation Technique	Holding Time
			Process Water			
BTEX	EPA 624/8240	1.0 μg/L	Borosilicate glass VOA[e] vials	3 x 40-mL	HCl to pH <2 @ 4°C	14 days
TPH (gasoline)	EPA Modified 5030/8015	0.5 mg/L	Borosilicate glass VOA vials	3 x 40-mL	HCl to pH <2 @ 4°C	14 days
TPH (diesel)	EPA 8015	0.5 mg/L	Amber glass bottle	1 to 2 L	Cool @ 4°C	14 days to extraction
Total iron	EPA 200.7	0.1 mg/L	Polyethylene bottle	500 mL	Nitric acid to pH <2, Cool @ 4°C	6 months
Alkalinity (carbonate)	EPA 310.1	20 mg/L	Polyethylene bottle	250 mL	NA	NA
			LNAPL			
BTEX	EPA Modified 8240	100 mg/L	Glass vial with Teflon™ system or lined cap	5 mL	Cool @ 4°C	14 days
Boiling-point distribution	GC/FID[f]	NA	Glass vial with Teflon™ system or lined cap	5 mL	Cool @ 4°C	14 days
Flash point	Method 1010	NA	Glass bottle	200 mL	NA	30 days
Percent moisture	ASTM D2216-17	NA	Glass or plastic bottle	500 mL	NA	NA
Total suspended solids	EPA 160.3		Glass or plastic bottle	500 mL	Cool @ 4°C	7 days

(a) MDL = method detection limit.
(b) ASTM = American Society for Testing and Materials.
(c) NA = not applicable.
(d) ppbv = parts per billion by volume.
(e) VOA = volatile organic analysis.
(f) GC/FID = gas chromatography/flame ionization detection.

and/or stable emulsions accumulate during the testing. If floating solids are not adequately removed, they may clog the conventional gravity OWS and the coalescer packings and eventually may be carried over to the surge storage tanks and treatment equipment downstream from the OWS. The stable emulsions may cause the water to appear milky and will increase the TPH concentration in the water, thus requiring additional treatment.

Two LNAPL samples should be collected from the fuel storage tank for BTEX, boiling-point distribution, flash point, moisture content, and suspended solids analyses. The results of the analyses may be used to determine if the LNAPL can be recycled or used for heating, or will need to be disposed of as waste.

3.6.2.6 Soil-Gas Monitoring

Monitoring of the soil-gas composition and pressure should be performed every 24 hours during the pilot tests using the monitoring points described in Section 3.3. The soil-gas composition data are used to evaluate how oxygen, carbon dioxide, and TPH concentrations in the soil gas vary with time. The pressure data may be used to determine the bioslurper vadose zone ROI. Finally, in conjunction with the results of the in situ respiration test, the soil-gas monitoring data may be used to estimate the mass of petroleum hydrocarbons removed from the vadose zone through biodegradation (in mg TPH/kg of soil/day). Using *n*-hexane as a model compound for TPH, a stoichiometric equation describing hydrocarbon degradation may be presented as follows:

$$C_6H_{14} + 9.5O_2 \rightarrow 6CO_2 + 7H_2O \qquad (3\text{-}1)$$

Based on this equation, approximately 3.5 g of oxygen is required, on a weight basis, for every 1 g of hydrocarbon consumed. Therefore, the hydrocarbon degradation rate is approximately 0.29 times the oxygen utilization rate on a weight basis.

3.6.2.7 LNAPL Thickness and Groundwater-Level Monitoring

During the bioslurper testing, the depth to groundwater and the LNAPL thickness may be measured in the monitoring wells adjacent to the extraction well using an in situ interface probe. If there are no existing wells, one or more wells may need to be constructed. Although monitoring the fluid levels in the wells is not absolutely necessary, useful data may be generated through this monitoring. Figure 3-9 depicts a typical arrangement of

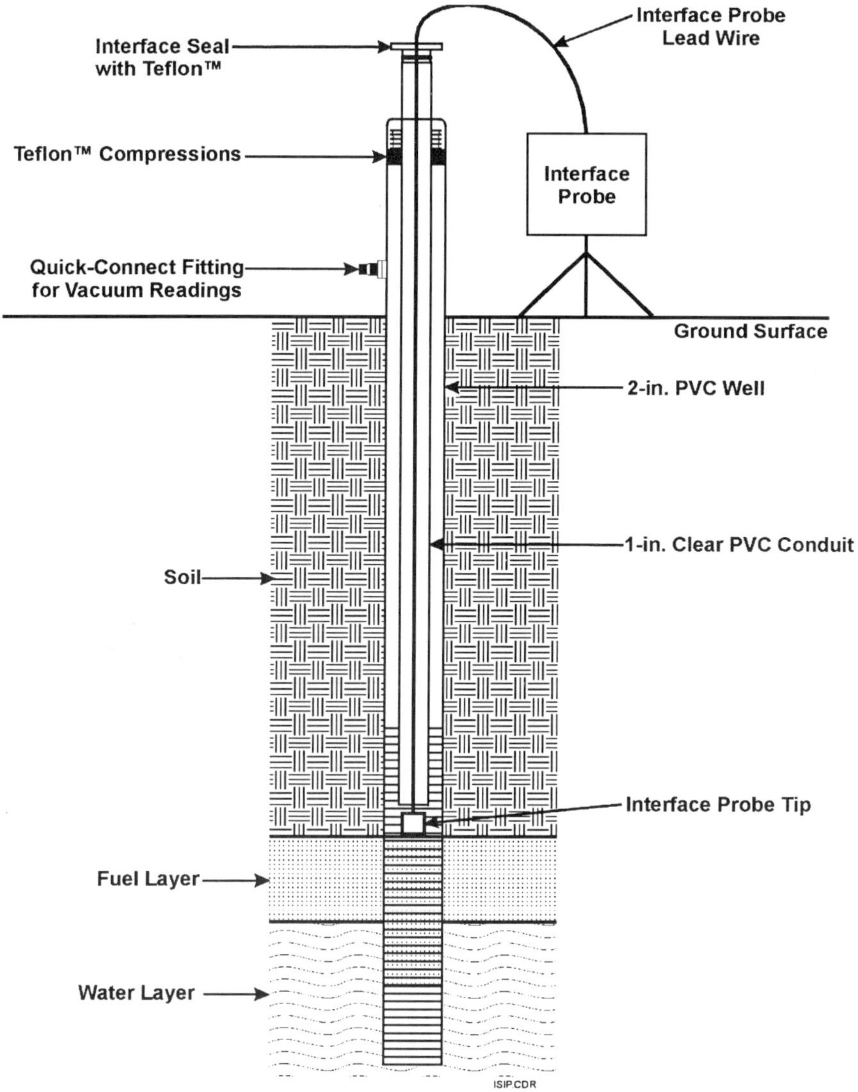

FIGURE 3-9. Diagram of the in situ interface probe setup

an in situ interface probe. This setup is ideal for measuring the LNAPL thickness and depth to groundwater at subsurface soil pressures created during the bioslurping test. The interface probe (placed in the well at the top of the LNAPL or groundwater and sealed tightly at the wellhead), is threaded through a section of 1-inch clear PVC tubing, which is fitted to a specialized well seal. The well seal has a Teflon™ gasket that seals the

PVC to the well casing. Teflon™ is self-lubricating, so the PVC pipe can be moved up and down in the well without short-circuiting to the atmosphere.

3.6.3 Pump Drawdown

Upon completion of the bioslurping test, a simulated pump drawdown test may be performed using the LRP and a single drop tube in the extraction well. The effect is the same as when using the single-pump method. The wellhead configuration is identical to that used during the simulated skimmer test, where the well casing is open to the atmosphere. The only difference between the simulated skimmer and the simulated drawdown configurations is the placement of the drop tube (Figure 3-10). During the

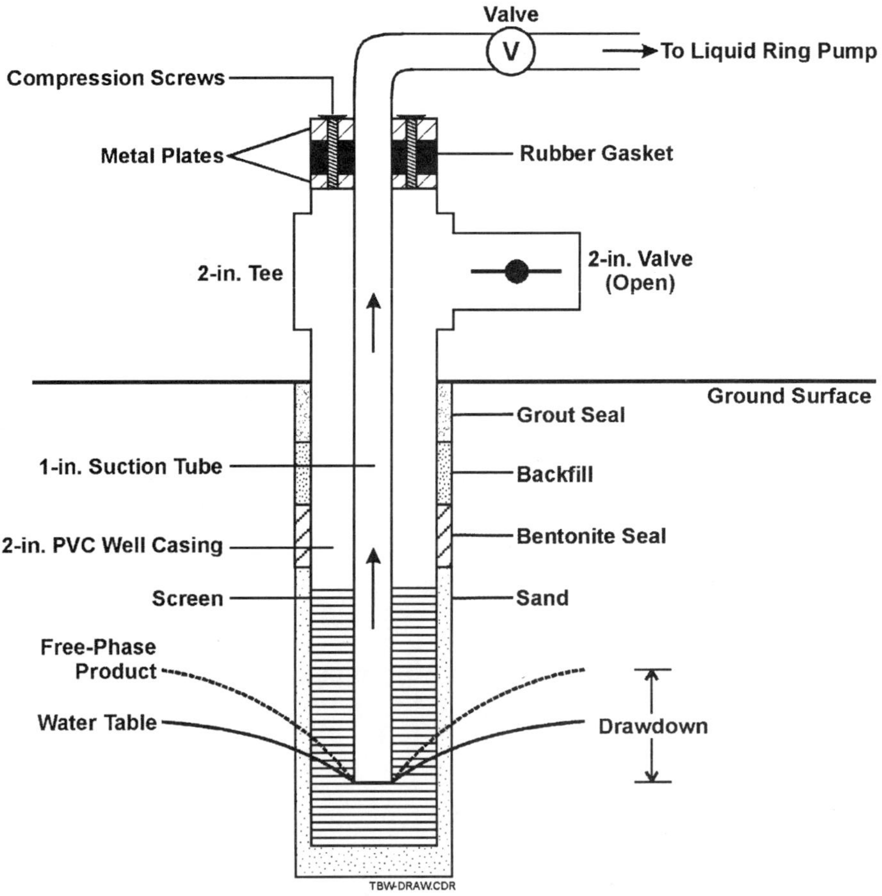

FIGURE 3-10. Configuration of drop tube and wellhead during simulated drawdown recovery test

simulated skimmer test, the drop tube is placed at the LNAPL/groundwater interface. However, during the simulated drawdown test, the drop tube is lowered below the measured static water level. In practice, the wellhead vacuum observed during the bioslurper test may be used to determine the placement of the drop tube for the simulated drawdown test. For example, if the bioslurper creates a vacuum of 15 inches H_2O at the wellhead, the drop tube should be placed at 15 inches below the static water table.

As with the simulated skimmer test, the LRP and OWS should be free of LNAPL and primed with water prior to the inception of the test. Readings on the totalizing flowmeters for water and LNAPL should be noted to provide a starting point for quantification of volumes recovered during the simulated drawdown test. Water and LNAPL recovery rates and TPH vapor concentrations should be monitored periodically over the duration of the test. Upon completion of the test (in about 24 hours), LNAPL is removed from the seal water tank and OWS, total volumes of recovered groundwater and LNAPL (including vapor phase) are determined, and depths to LNAPL and groundwater are measured.

3.6.4 Simulated Soil Vapor Extraction

The stack gas discharged from the LRP is monitored during the simulated skimmer, bioslurper, and simulated drawdown tests. Petroleum hydrocarbons in the vapor discharge during the simulated skimmer test most likely are from the volatilization of recovered LNAPL in the seal water tank. The petroleum hydrocarbons in the stack gas during the bioslurper test are from the recovered soil gas as well as from the volatilization of recovered LNAPL in the seal water tank. If TPH concentrations in the vapor discharge during the bioslurper test are significantly higher than during the simulated skimmer test, a simulated SVE test may be performed. The TPH concentrations in the stack gas also may be compared to those in the soil gas. If concentrations in the stack gas during the bioslurper test are comparable to those in the soil gas, an SVE test should be conducted.

The simulated SVE test is performed using the LRP to create a vacuum in the vadose zone. The drop tube should be raised above the static liquid level in the well to prevent recovery of fuel or water by the pump, allowing the pump to recover only soil vapor. The well should be closed to the atmosphere, as during the bioslurper test, to allow the creation of a vacuum in the well and the surrounding vadose zone. Liquid levels in the extraction well should not be allowed to rise above the screened interval. In order to prevent elevated TPH concentrations in the stack-gas discharge due to volatilization, any free product present in the LRP and seal water tank must be removed prior to the inception of the test. The OWS may be ignored, as no liquids will be recovered during this test. Vapor discharge concentrations and flowrates should be measured during the test to calculate the mass of TPH removed in the vapor phase.

3.7 Data Reduction and Results Interpretation

Figure 3-11 presents a decision tree for determining if bioslurping is an appropriate technology for full-scale implementation at an LNAPL-contaminated site. The data collected from baildown tests performed during the site characterization provide an indication of whether free product at the site is mobile under passive conditions. If the recovery rates during the baildown tests are less than 0.1 gpd, LNAPL recovery using bioslurping, skimming, or drawdown may not be economically feasible. Therefore, it is recommended that other technologies, such as bioventing or SVE be investigated for site remediation. Data and information collected during the pilot-scale tests should be evaluated to (1) determine the feasibility of the LNAPL source removal and vadose zone remediation, and (2) develop the approach for construction and implementation of a full-scale bioslurper system. Sections 3.7.1 through 3.7.4 present information on how the pilot test data can be used to develop a full-scale bioslurper system. These sections describe the evaluation of the data from operating a bioslurper system at a single extraction well. The bioslurper may be used to extract from multiple wells simultaneously to generate data that will assist in the full-scale implementation of bioslurping. These data will be more representative of the flow regimes (LNAPL, water, and stack gas) during full-scale operation, and will aid in determining the size of the pump and the selection of stack-gas and discharge water treatment systems.

3.7.1 LNAPL Thickness Monitoring

Monitoring the LNAPL thickness throughout the pilot test, as described in Section 3.6.2.7, may provide information regarding the capture zone of the bioslurper extraction well. If the monitoring well is within the ROI of the extraction well, a small reduction in the apparent LNAPL thickness in the monitoring well may be observed. Care must be taken when interpreting the fluid level monitoring data. A change in the thickness of LNAPL in the monitoring well may be the result of a fluctuation in the water table caused by precipitation or drawdown of the water table caused by improper placement of the bioslurper drop tube.

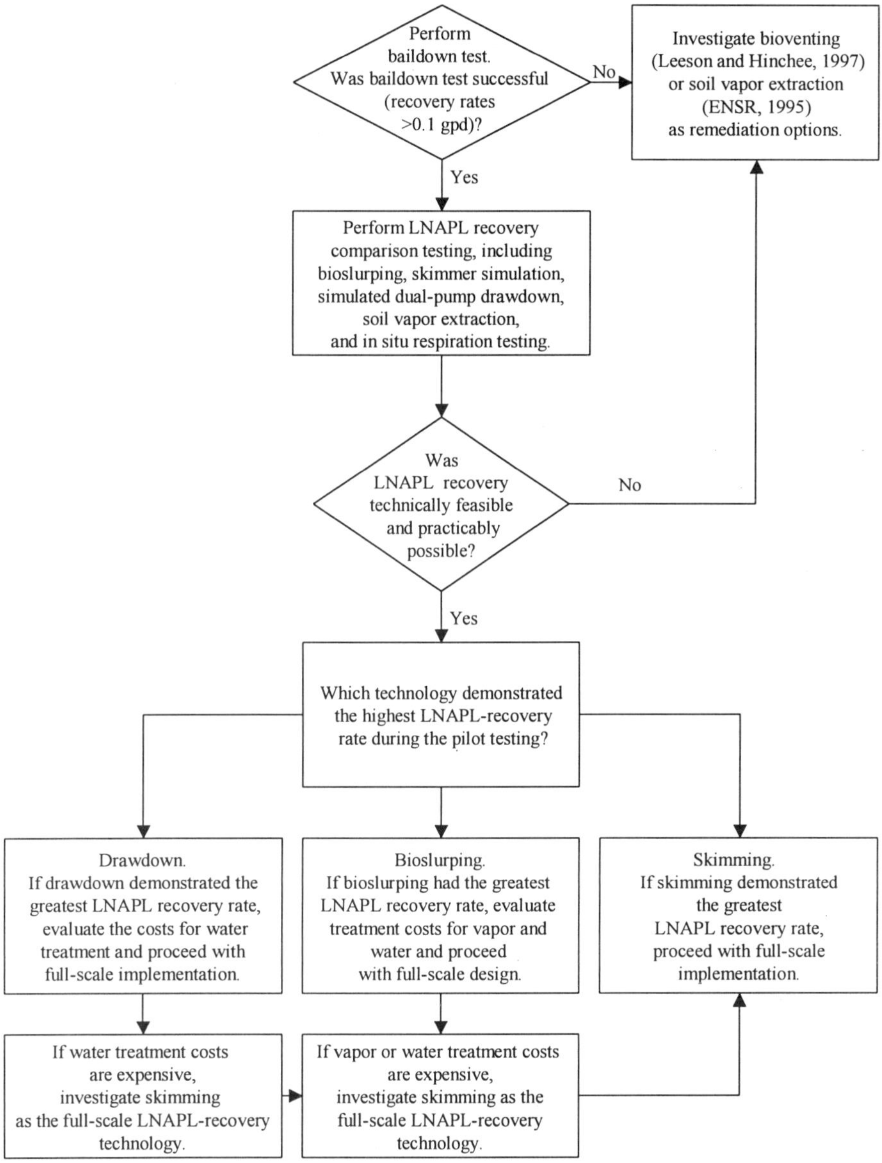

FIGURE 3-11. Decision tree for selecting an LNAPL-recovery technology

3.7.2 LNAPL Radius of Influence

Traditionally, the ROI of the bioslurper for the LNAPL zone and full-scale design has been estimated based on the vadose zone ROI determined during the bioslurper pilot tests. However, in most cases, the flow behavior of the LNAPL is likely to be different from that of the soil gas in the unsaturated zone due to complicating factors related to multiphase fluid movement. Therefore, the ROI determined from the vadose zone of influence may overestimate the ROI in the LNAPL layer. The design of full-scale bioslurper systems based on these ROIs may result in gaps in the LNAPL capture zones in the field.

A more rigorous estimation of the LNAPL ROI may be made by combining the field data collected during site characterization and pilot-scale testing with multiple-phase computer simulations. The numerical simulations can incorporate the hydrogeologic data, LNAPL distribution, LNAPL type, and the vacuum rates into a single model. The calculations involve solution of highly nonlinear equations resulting from changes in relative permeability and saturation to calculate the saturation, pressure distribution, and the velocity field for each phase (soil gas, LNAPL, and groundwater). The models also can estimate the recovery of each phase. One major advantage is that they can be used to conduct sensitivity analyses to evaluate the effects of parameter uncertainty on the ROI and overall system performance. Although three-dimensional codes are desirable, most currently available codes are two-dimensional (plan view or cross-sectional view). An example of a plan view code is MOVER (DAEM, 1997), a multiphase, finite-element code specifically written for bioslurping processes.

The simulation of a vacuum-enhanced recovery system at a typical site requires data on LNAPL spatial distribution; aquifer geometry; soil properties such as permeability, porosity, van Genuchten parameters (alpha and n), and residual saturation; and fluid surface tension, viscosity, and density. Accurate surveys of monitoring well locations and elevations are required. It is useful to obtain data for observed oil and water levels from several measurement events to account for seasonal variations in water levels. Similarly, soil property measurements at several locations provide an estimate of the variability of the properties and improve confidence in the model results. Further adjustment in the soil properties may be made during calibration when an attempt is made to match field pilot-scale observations, such as the soil-gas vacuum, with model results.

As a first step in the MOVER plan view simulations, the monitoring well data are used to determine the actual saturation distribution of LNAPL in the soil and to estimate the total initial LNAPL volume. The accuracy of the estimate depends on the reliability of the observed LNAPL distribution, the soil property variations (heterogeneity), and the model setup. The process of bioslurping is simulated by applying a constant vacuum condition with pressure expressed in inches of water vacuum at the bioslurper well location in the model domain. This pressure gradient is the driving force for vacuum-enhanced recovery, and the movement of each phase is governed by its piezometric head gradient.

As an example of the model use for estimating the ROI, the MOVER code was used to simulate pilot-scale bioslurping at an LNAPL site. The formation was assumed to be composed of uniform, medium-sized sand with the water table at 15 ft bgs, the LNAPL thickness varying from 0 to 2 ft, and a groundwater velocity of about 20 ft/year. The model was run for a single well with a vacuum head equivalent to 5.6 ft of water applied for about one day and the suction tube placed at the oil/water interface. The output-simulated velocity vectors for the oil, water, and soil-gas phases were resolved and plotted to evaluate the zone of influence around the bioslurper well in each phase. Simulated flow direction vectors and the zones of influence for the oil phase and air phase are shown in Figures 3-12 A and B, respectively. Based on Figure 3-12A, the LNAPL ROI is about 27 ft in the east-west direction and about 40 ft in the north-south direction with the asymmetrical shape probably due to natural groundwater and/or LNAPL flow. The ROI based on the air velocity appears to extend throughout the model area and is much larger than that determined for the oil phase. The effective ROI in the vadose zone may have to be based on the arbitrary cutoff point for pressure, such as 0.1 inch vacuum. In the example site given here, the field ROI based on the pilot-test soil-gas pressure was 45 ft. In general, it appears from the example model that the LNAPL zone ROI is much smaller than the vadose zone ROI determined from either the model or the field data.

The ROI for the bioslurper should be based on the flow in the LNAPL phase rather than in the soil-gas phase. Although computer simulations can be used in conjunction with the pilot-scale tests to improve the ROI calculations, these require some amount of specialized expertise not available at many Naval facilities or companies. It may be cost-effective to develop some generalized rule-of-thumb–type guidance that relates the observed field parameters to the actual LNAPL ROI without conducting complex simulations.

3.7.3 LNAPL Extraction Tests

The results from the LNAPL extraction pilot test, which compares the different extraction technologies, may be used to determine the most cost-effective technology or combination of technologies for remediation. If bioslurping is selected for site remediation, the information generated by the LNAPL extraction test may be used to design a full-scale system. Some of the most useful pieces of data are:

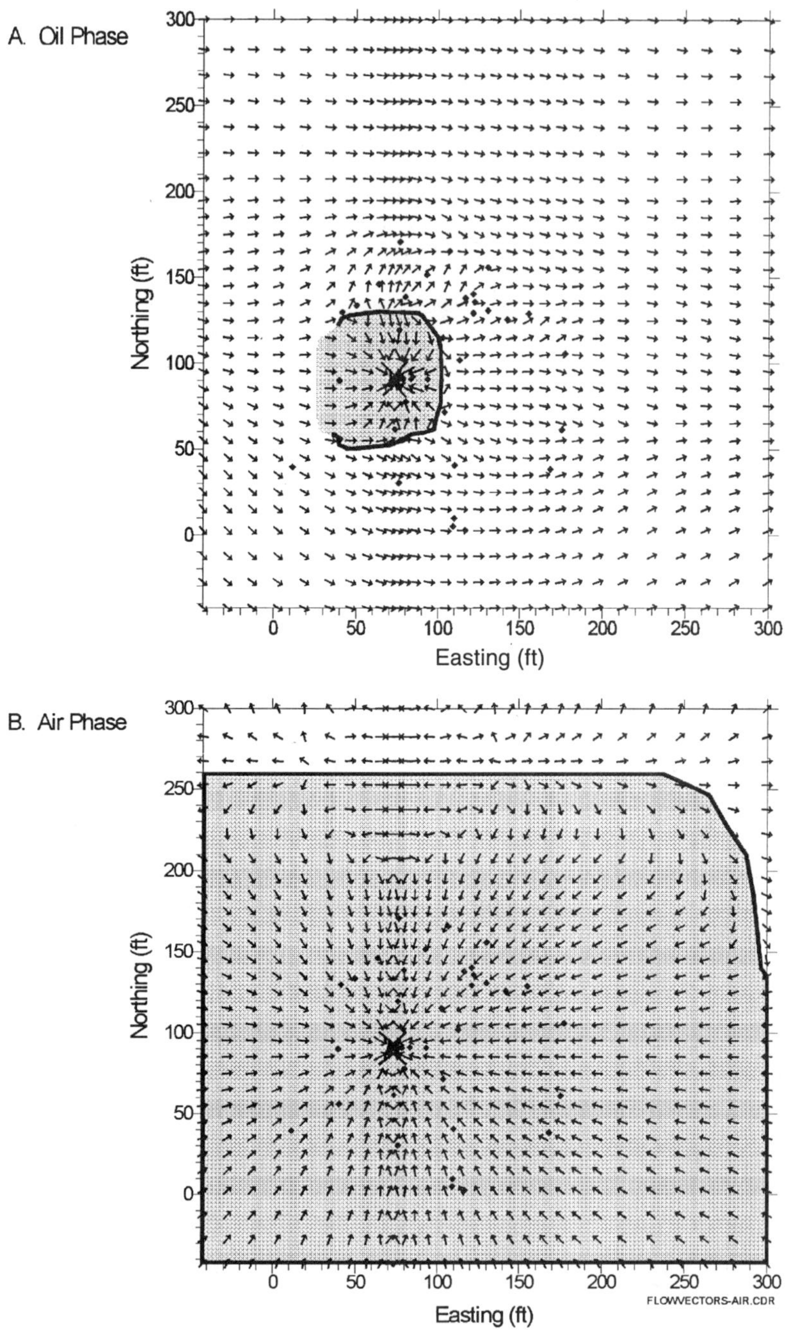

FIGURE 3-12. Simulated flow vectors for (A) the oil phase and (B) the air phase

- LNAPL recovery rate
- Groundwater recovery rate and process water composition
- Stack-gas discharge flowrate and composition
- Formation of floating solids and stable emulsions.

This information is required for sizing the full-scale system (e.g., pump capacities and sizes of OWS and holding tank) and determining the requirement for oil/water separation and the need for stack-gas or process water effluent treatment. Data generated during the pilot studies also may be used for discharge permit applications, if permitting is required. If the stack gas and/or process water effluent requires treatment, composition and flowrate data must be used to obtain sizing information, performance data, and cost information from various vendors and component suppliers for the full-scale treatment units.

The LNAPL recovery rate may be used to determine the most cost-effective technology for removing LNAPL from the site. In general, the technology that sustains the greatest LNAPL recovery rates during the pilot test should be considered for full-scale remediation of the site. However, prior to selecting and implementing a specific technology, other factors should be evaluated, such as the groundwater recovery rate, TPH concentration in discharge water, stack-gas discharge rate, TPH concentration in the stack gas, and biodegradation rate in the vadose zone. If accurate LNAPL volume estimates are provided, the LNAPL recovery rate may be used to estimate the O&M time required to remove the LNAPL from the site. The O&M time and costs and effluent discharge treatment costs are important when deciding which technology is most appropriate for a particular site.

If low LNAPL recovery rates are measured during each of the LNAPL recovery configurations, if may not be possible to economically recover the LNAPL. In such cases, it is recommended that a non-LNAPL recovery technology such as bioventing or SVE be investigated for possible use in full-scale site remediation, and that bailing be used to remove the LNAPL, if necessary. The simulated SVE test and the in situ respiration test conducted during the LNAPL recovery pilot test provide an indication of the potential to use these technologies for site remediation. However, before eliminating the use of any of the technologies, seasonal variations in the contaminant profile should be considered. For example, if fuel is trapped beneath the water table or floating above the screened interval of the extraction well during a period of high precipitation, drawdown may be the only LNAPL recovery technology capable of removing significant quantities of LNAPL. If such a situation is known to exist, the pilot test should be repeated during a drier period.

The characteristics and volume of the wastestreams generated for each technology should be considered when performing the cost analysis for site remediation. The mass discharge rate of TPH in the discharge water and off-gas may eliminate the application of a specific technology if the disposal/treatment costs are expensive for the water or vapor. If the carbonate concentrations are high in the process water or in the groundwater, there is a potential for scale formation in the LRP during bioslurper operations. In most cases, the scale may be removed by adding a dilute acid, such as vinegar, to the LRP during operation.

3.7.4 Soil-Gas Monitoring and Shutdown Respiration Testing

During the bioslurper pilot test, the in situ soil-gas composition is measured at the monitoring points. Along with the information on the vadose zone ROI, the soil-gas data may be used to determine whether oxygen and/or carbon dioxide levels have increased. Increased oxygen levels might indicate bioslurper-mediated aeration, and increased CO_2 levels may indicate enhanced microbial activity in the area of influence.

When the bioslurper extraction tests are complete, a shutdown respiration test may be performed to evaluate the oxygen utilization rate and the biodegradation rate in the vadose zone. The soil-gas data collected over time (after shutdown of the bioslurper test) may be used to establish the oxygen utilization rate and the carbon dioxide evolution rate. High oxygen utilization rates (>1% per day) indicate improved microbial activity by the bioslurper-mediated aeration. If the oxygen utilization rates are low and significant amounts of contaminants are present, factors such as high clay content, low or very high moisture content, low soil pH, nutrient limitation, and/or contaminant levels toxic to microorganisms may be limiting aeration and/or biodegradation rates. As such, site-specific variables affecting microbial degradation must be identified to determine whether the conditions can be improved to implement enhanced bioremediation.

3.8 References

DAEM, see Draper Aden Environmental Modeling, Inc.

Draper Aden Environmental Modeling, Inc. 1997. OILVOL: Oil Volume Estimator. Technical Documentation and User's Guide. Blacksburg, VA. 11 p.

ENSR. 1995. *Soil Vapor Extraction and Bioventing*. EM1110-1-4001. Prepared by ENSR under contract to U.S. Army Corps of Engineers. November.

Leeson, A., J.A. Kittel, R.E. Hinchee, R.N. Miller, P.E. Haas, and R.E. Hoeppel. 1995. "Test Plan and Technical Protocol for Bioslurping." In R.E. Hinchee, J.A. Kittel, and H.J. Reisinger (Eds.), *Applied Bioremediation of Petroleum Hydrocarbons*. pp. 335-347.

Leeson, A., and R.E. Hinchee. 1997. *Soil Bioventing Principles and Practices*. CRC Press Lewis Publishers, Boca Raton, FL. pp. 244.

4

Full-Scale Bioslurper Design and Construction

Chapter 4 presents the methods used to design a full-scale bioslurper system using the data obtained during the characterization and pilot-scale test (Chapters 2 and 3 of this book, respectively). Figure 4-1 presents a stepwise approach for designing a full-scale bioslurper system. It is important to follow the flow in Figure 4-1 to simplify the design procedure. If the flow is not followed, full-scale pump sizing and the requirements for off-gas and process water treatment may be difficult to establish. The formation of emulsions and floating solids has been a problem observed at many bioslurping sites and is primarily attributed to the blending action of the LRP. Section 4.4 presents effective techniques for removing emulsified fuel and floating solids from the liquid stream. Additionally, separation using a knockout tank and in situ OWS are presented in Section 4.4. Such separation techniques remove the LNAPL from the liquid stream prior to entering the pump, thus reducing the formation of these byproducts. Process water and stack-gas treatment options are presented in Sections 4.5 and 4.6, respectively.

4.1 Extraction Well Spacing and Design

Placing wells with proper spacing is important to ensure optimal LNAPL recovery and vadose zone remediation. The design, placement, and installation of multiple bioslurper wells are some of the most critical elements in achieving successful implementation of the bioslurper technology. The desired well spacing is determined by the ROI estimated during the pilot test and modeling. Wells should be spaced in a configuration such that the distance between any two adjacent wells does not exceed the ROI. Optimization of the number and spacing of extraction wells would minimize both capital and operating costs.

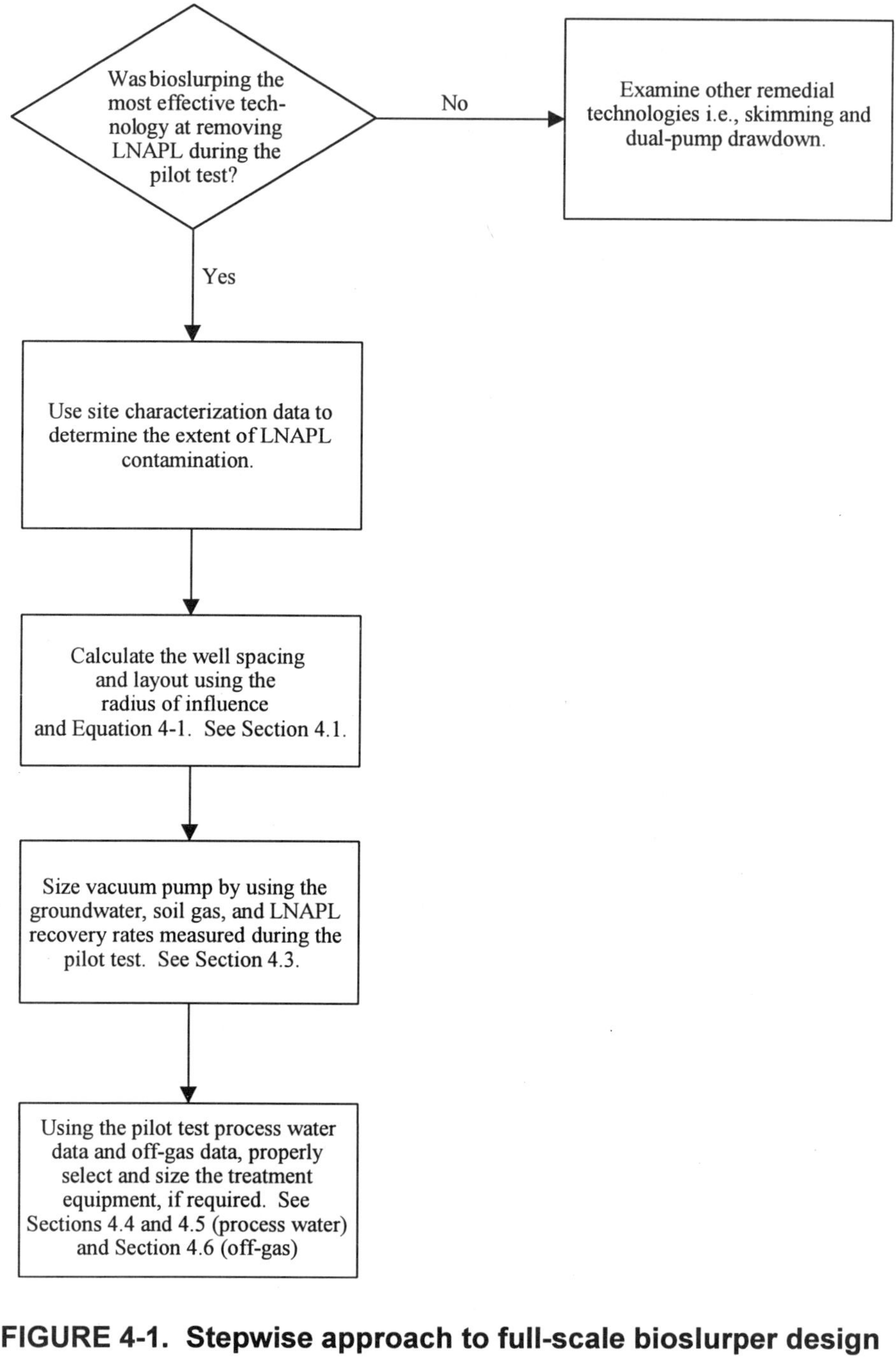

FIGURE 4-1. Stepwise approach to full-scale bioslurper design

The extraction wells should be installed equidistant from one another in triangular arrays, as depicted in Figure 4-2. With arrays of this type, there will be some overlap of the area of influence for each well. Using the ROI from the pilot test, the spacing between the wells can be determined from the following equation:

$$L = 2r\cos(30) = 1.732r \qquad (4\text{-}1)$$

where: L = the distance between wells
r = the ROI measured during the pilot test.

Using Equation 4-1, there is minimal overlap of the radii of influence from the wells. If greater overlap is desired, a multiplier may be incorporated into the equation to provide an acceptable overlap.

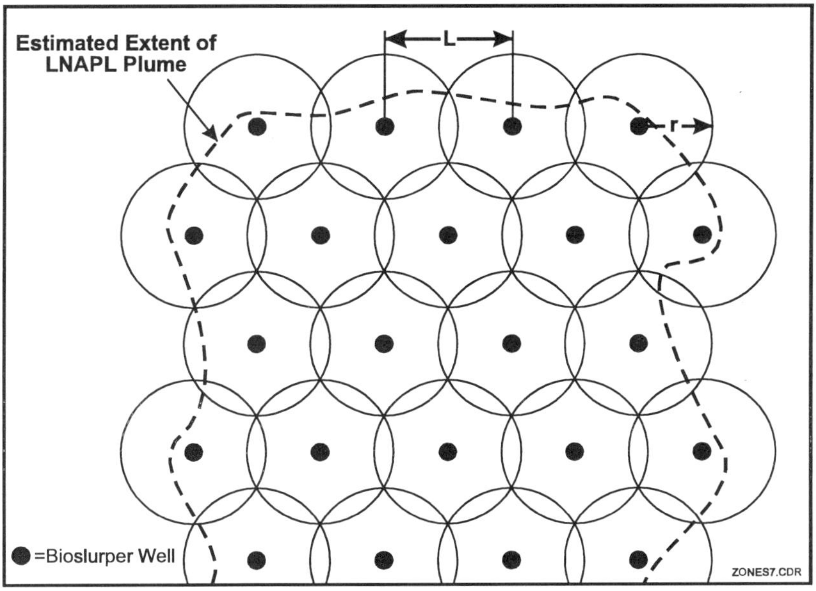

FIGURE 4-2. Recommended extraction well layout

4.2 Drop Tube and Extraction Manifold Sizing

The bioslurper drop tube is designed to ensure that the superficial velocity of the soil gas (V_{SG}) and the superficial velocity of the liquid (V_{SL})

are low enough to maintain a pulsing or annular-type flow in the vertical pipe. Typically, a large vacuum is not required to lift fluids from the water table to the ground surface. Rather, the liquids generally are lifted through air entrainment. In general, a V_{SG} of >1.0 ft/sec would be adequate to maintain such a flow (Govier and Aziz, 1972). Meanwhile, the soil-gas flow velocity would be low enough (i.e., <500 ft/sec) to avoid choked flow conditions. Figure 4-3 displays the flow patterns at these superficial liquid and gas velocities. Testing of the bioslurper has demonstrated that the suction lift action is not very sensitive to operating conditions and that stable operation is easily maintained over a wide range of conditions.

Piping for the manifold must be selected so that the liquid velocity decreases sufficiently to avoid slug flow in horizontal piping. Pressure losses due to friction must be taken into account when designing a manifold system. Data for frictional loss in pipes can be found in engineering manuals or may be obtained from the pipe manufacturer. Again, the soil-gas velocity should be low enough (V_{SG} of <500 ft/sec) to avoid approaching choked flow conditions. The sizes of manifold piping may increase near the LRP as more wells are connected to the manifold. As much as possible, the manifold piping should be placed to avoid low points and should be sloped toward the LRP.

4.3 Liquid Ring Pump

From experience at many bioslurper sites, the LRP is an economical and effective system for producing a liquid-extraction vacuum. LRPs generally are very efficient and able to produce vacuum levels greater than 25 inches Hg. While a large vacuum may not be required to lift fluids to the surface (as stated in Section 4.2), greater vacuum within the extraction well typically will increase the LNAPL recovery. Twin-lobe rotary blowers have been used as an alternative to the LRPs. Unlike the LRP, the rotary blower will not tolerate any water. If a rotary blower is used, a liquid/air separator must be installed prior to the blower such that liquid never enters the blower. The rotary lobe blowers prevent the mixing of groundwater and LNAPL in the extraction process that occurs due to the shearing action in the LRP. From experience, this benefit may be minimal at some sites, because mixing of the groundwater and LNAPL may occur prior to entering the pump.

The rule of thumb used in the past for sizing LRPs at bioslurping sites is to allow between one-third and two-thirds horsepower for each extraction well. However, the primary factor in sizing the LRP for a full-scale

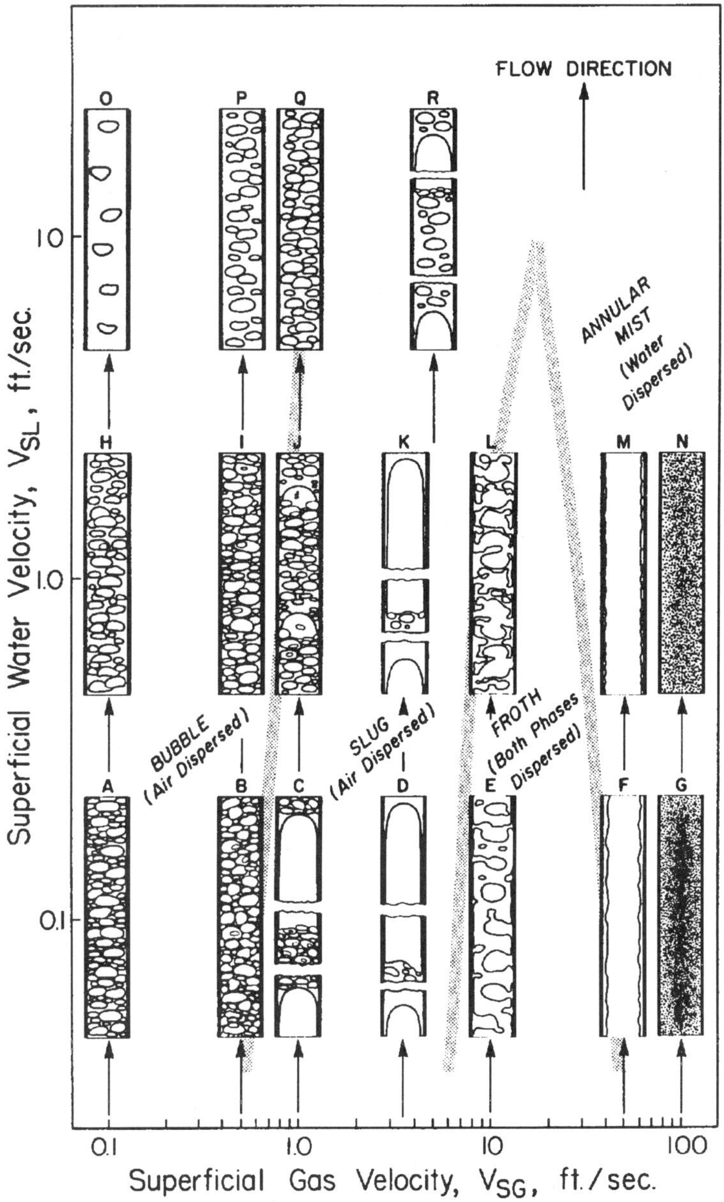

FIGURE 4-3. Flow patterns at superficial liquid (V_{SL}) and gas (V_{SG}) flow velocities (Reproduced from Govier and Aziz, *The Flow of Complex Mixtures in Pipes*, © 1972. Used with permission of Krieger Publishing Co., Inc., Melbourne, FL.)

bioslurper system is the volumetric flowrate of soil gas to extract fluids from the wells via the drop tube.

The LNAPL, water, and air flowrates measured during the pilot test are multiplied by the number of wells required to cover the LNAPL plume to give the flowrate for each component. The only method to produce the same ROI during the full-scale effort as was measured during the pilot effort is to extract groundwater, soil gas, LNAPL at the same rate (per well) and ratio during both phases of operation. If the vacuum pump was operated at only one setting, the above calculation typically will estimate the maximum size for the LRP. However, if liquid and soil-gas extraction rates were reduced during the pilot test to evaluate the system operation at lower-flow conditions, the data generated during this evaluation may be used to provide the minimum size of the LRP. LNAPL, water, and stack-gas flowrates are multiplied by the number of wells required to cover the LNAPL plume just as is done to estimate the maximum pump size.

The required pump size can be determined by conducting multiple-well extraction testing. To conduct this type of test, the system initially extracts from a single well. Throughout the testing the fuel, groundwater, and soil-gas recovery rates; wellhead vacuum; pump vacuum; pump temperature; and stack-gas TPH concentrations are monitored. After the water and soil-gas recovery rates equilibrate, a second well is added to the system. This procedure is repeated as five to ten wells are connected to the LRP. The total flowrate for fuel, water, and vapor is then divided by the number of wells connected to the pump to calculate the average flowrate from each well.

Multiple-well testing was performed at a naval facility in Florida and the results are provided as an example. A total of five wells were connected to the system by the end of the test. Throughout the testing sequence the overall soil-gas flowrate increased. A soil-gas flowrate of 61 standard cubic feet per minute (scfm) (61% of the theoretical capacity) was observed while extracting from five wells. Although the overall soil-gas flowrate increased, the soil-gas flowrate from the individual wells decreased. Figure 4-4 illustrates the change in the average soil-gas flowrate per extraction well as additional wells were added to the bioslurper system. The flowrate was as high as 23 scfm when extracting from one well; flowrates dropped to about 11 scfm/well while extracting from five wells.

To estimate the total soil-gas flowrate anticipated during full-scale operation, the flowrate data displayed in Figure 4-4 were fitted with an exponential curve having the following equation:

$$Y = 22.4*X^{-0.424} \tag{4-3}$$

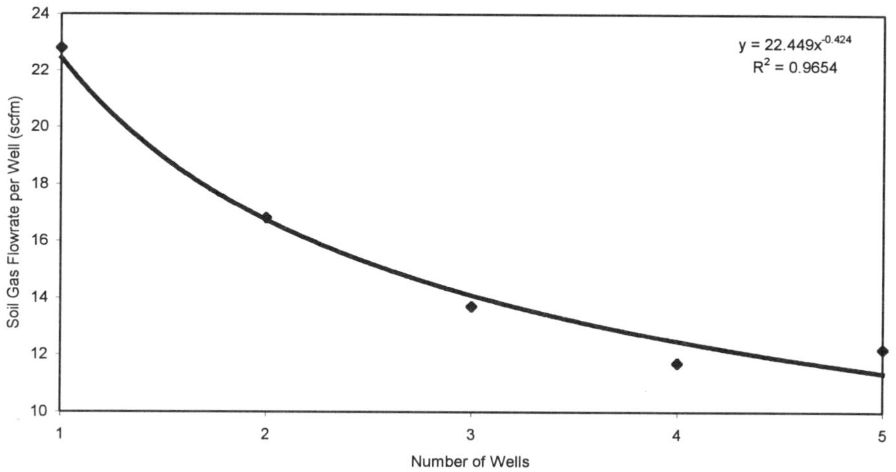

FIGURE 4-4. Effect of multiple wells on soil-gas extraction rate

where: Y = soil-gas flowrate/well
X = number of extraction wells.

It was assumed that 17 wells would be utilized during the full-scale operation, and the resulting soil-gas flowrate would be 6.8 scfm per well. The total flowrate would then be 116 scfm or 218 cfm at 14 inches Hg vacuum. A vacuum of 14 inches Hg was used to convert scfm to cfm because the pilot-scale test data (while extracting from five wells) indicated that sufficient vacuum could be generated at the wells while operating the LRP at 14 inches Hg. Based on the literature provided by the pump manufacturer, a 15-hp pump was selected.

Pumps in the capacity range often used in bioslurping systems (1.5 to 20 hp) are available with 220-V single-phase motors or 240/480-V three-phase motors. The single-phase versions generally are preferred for pumps in the 1.5- to 5-hp size range. The 240/480-V three-phase version is preferred for pumps over 5 hp due to its efficiency and lower current draw. The designer should determine the site's voltage supply capacity before selecting the vacuum pump size.

If the required horsepower exceeds 15, it may be more efficient to use more than one pump. Using several lower-horsepower pumps avoids high current flow that can complicate power supply systems, increases the flexibility of the system, and allows at least partial operation if one pump is out of service.

4.4 Oil/Water Separation

LNAPL, groundwater, and soil gas are extracted through the same extraction manifold during a typical bioslurping operation. Therefore, some type of OWS is required to separate LNAPL from the liquid stream, especially if floating solids and stable oil/water emulsions are present in the stream. At some bioslurper sites, the formation of floating solids and stable emulsions presents a significant problem with the operation of the technology. The floating solids observed at most bioslurper sites appear as a frothy mass floating between the extracted LNAPL and groundwater. The oil/water emulsions are suspended droplets of petroleum hydrocarbons in water that give the process water from the bioslurper a milky appearance. The emulsions may be produced during the mixing action of the LRP or the slurping action in extraction wells. The floating solids are formed from the mixture of the extracted LNAPL, soil gas, groundwater, and sediment captured during the extraction process. In general, the floating solids and milky oil/water emulsions are relatively stable, thus preventing the successful separation of LNAPL and groundwater by a conventional gravity OWS alone. If an OWS is properly designed and sized to handle the oil/water separation, it can be effective in separating the floating solids. Several techniques have been used to reduce the formation of the emulsions and/or remove the emulsions from the aqueous stream once they have formed. These techniques are discussed in Sections 4.4.1 through 4.4.4.

4.4.1 Separation Using a Knockout Tank

As shown on Figure 4-5, before entering the LRP, the mixture of LNAPL, groundwater, and soil gas from an extraction manifold may be separated using a knockout tank assembly consisting of a vacuum-resistant tank with a piping system. The upper section of the piping removes soil gas, while the lower section removes groundwater. The liquid level in the tank is kept constant by removing soil gas and groundwater at the same rate at which they enter the tank. The LNAPL floating on the surface of the groundwater accumulates in the tank until it reaches a preset level where gravity causes it to drain to a fuel storage tank. The fuel storage tank is kept under vacuum to allow the LNAPL to drain properly. The connection between the knockout tank and the manifold must be located above the static fluid level in the tank, which is determined by the location of the tee fitting.

Initial testing of the knockout tank to prevent the formation of the floating solids and stable emulsions had limited success, which was attributed to

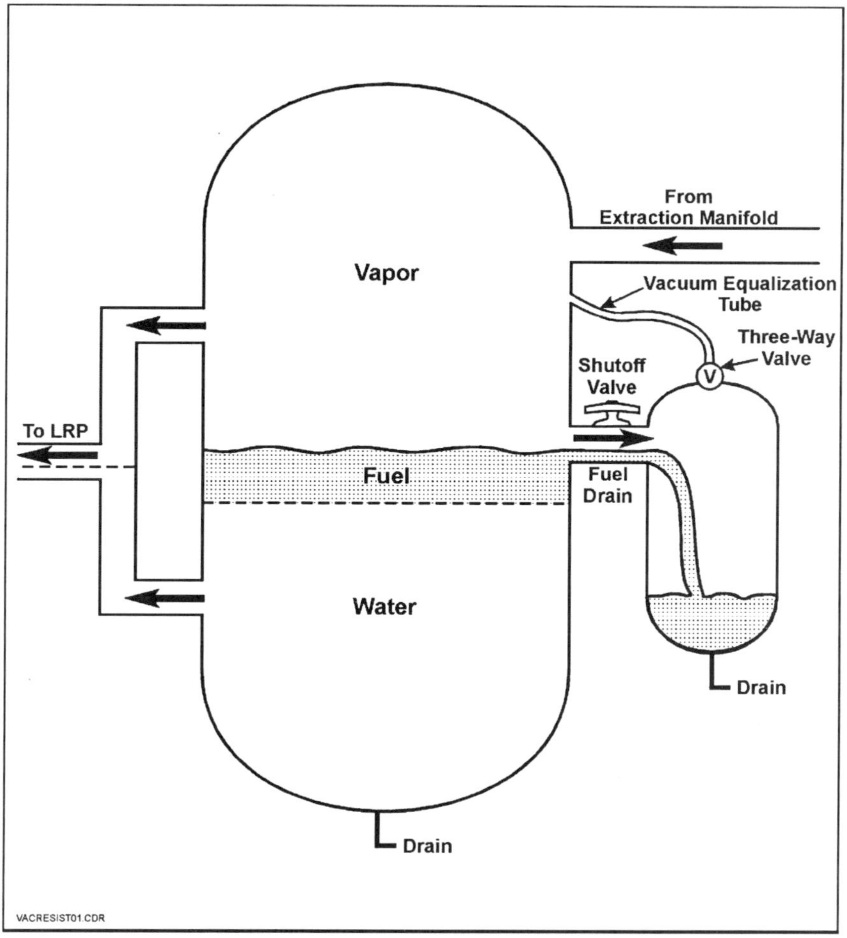

FIGURE 4-5. Vacuum-resistant separator

the complexity of the liquid level adjustment valves on the knockout tank. The complex nature of the valves and the variable groundwater recovery rates prevented proper setting of the liquid level in the tank, thus preventing satisfactory separation from occurring. Following these early attempts, the knockout tank assembly was modified to allow the fluid levels to "self-equalize." This modification has significantly improved the separation capability and minimized the O&M requirements. Results of tests at sites with modified knockout tank assemblies indicate that the formation of floating solids can be prevented and TPH concentrations in process water can be reduced by 72%.

4.4.2 In-Well Oil/Water Separation

As mentioned previously, floating solids and emulsions may form in the manifold before they are subject to mixing in the LRP. The potential for the production of these solids and emulsions may be significantly reduced if LNAPL and groundwater can be separated before extraction. In-well oil/water separation may provide an effective means to achieve this goal. One dual-drop tube design for in-well separation is displayed in Figure 4-6. A single aboveground vacuum pump is used to enhance the migration of LNAPL to the extraction well, similar to the original single-drop tube design. However, with the dual-drop tube design, LNAPL and groundwater are extracted in two separate streams through two separate drop tubes. The two drop tubes (1-inch PVC and ¼-inch tubing) are placed in the extraction well with the end of the 1-inch drop tube placed at the LNAPL/water interface and the end of the ¼-inch drop tube placed approximately one-half inch above the interface. Meanwhile, the end of the

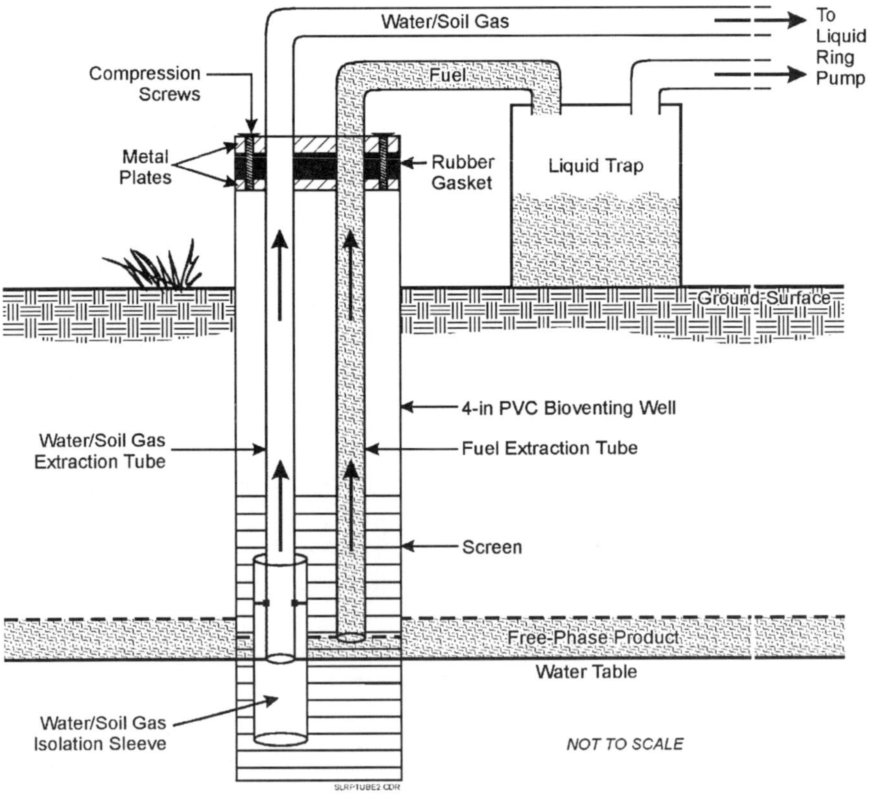

FIGURE 4-6. Dual-drop tube design for in-well separation

1-inch drop tube is shielded with a 2-inch-diameter PVC pipe, extending both 1 ft above and 1 ft below the end of the drop tube. The section of the 2-inch pipe prevents LNAPL from entering the deeper drop tube, while allowing soil gas and groundwater to be extracted. The 1-inch drop tube enhances the migration of LNAPL to the extraction well without producing a cone of depression. The ¼-inch drop tube extracts LNAPL from the well and transports it to a liquid trap before it reaches the LRP.

The dual-drop tube design and several other in-well separation configurations are being tested. The results of the dual-drop tube tests indicate as much as a 99% decrease in TPH in the process water while preventing the formation of floating solids. It is recommended, however, that the new drop tube design not be incorporated into the full-scale design until more thorough testing has been completed at sites with differing fuel contamination and soil types.

4.4.3 Separation Through Physical Removal

Physical removal of the floating solids using tanks equipped with fibrous filters has been successful during short-term projects. However, the operation and maintenance of these filter tanks can be time-intensive if large amounts of the floating solids are produced, because the filters quickly become clogged and must be replaced. Therefore, it is recommended that the filter tanks (or equivalent) not be used for full-scale bioslurper implementation.

4.4.4 Separation Using Coalescers

Coalescers are devices often added to simple gravity separators to enhance the performance of oil/water separation. Coalescers bring together small oil droplets to form larger droplets that can be separated more easily.

Oil droplets in water will attach preferentially to surfaces, particularly hydrophobic surfaces, and the oil phase will then wet the surface of the coalescing media to form a film. Additional oil droplets will coalesce onto the film until the oil droplets are large enough to break away from the film and rise to the phase layer interface. Although all emulsions are unstable, many are sufficiently stable to separate efficiently in a gravity separation tank. With a coalescing device, small droplets that by themselves would not separate by gravity in a reasonable time will coalesce into large droplets that separate more efficiently.

Typical coalescing media are composed of plates, beads, meshes, screens, and membranes made from oleophilic materials such as polypropylene, nylon, polytetrafluorocarbon, glass, and glass-treated materials. The best coalescers consist of a composite device with stages of increasing particle size surface area and with a final stage that enables the release of oil droplets from the solid media. Coalescers can separate oil droplets that are 10 μm or smaller in diameter. Droplets as small as 1 μm can be removed by coalescing devices. Coalesced droplets typically leave the coalescing device in droplet sizes ranging from 150 to 1,000 μm.

The velocity of the fluid flowing through the coalescing device is a key operating parameter. The velocity must be low enough that the droplets can grow sufficiently before being swept off the coalescer media. Commercial designs typically operate in the range of 1 to 10 ft/min. Separators with coalescing devices usually require a greater degree of maintenance and monitoring during operation than simple gravity separation systems.

Coalescing devices are ineffective in removing chemically stabilized emulsions that have a decreased interfacial tension and are less likely to be coalesced. Fine, suspended solids also may limit the effectiveness of a coalescing device because the solids often adhere to the surface of the coalescer and eventually clog the coalescer.

Coalescers are commercial off-the-shelf technologies and are available in many different forms. Slant-rib coalescing separators manufactured by Great Lakes Environmental® Products (see Figure 4-7) have been used at several bioslurper sites and provide satisfactory oil/water separation. The floating solids are removed by spilling over an oil weir into the oil reservoir.

4.5 Groundwater Effluent Treatment

The process water from an OWS must be treated and/or disposed of in a suitable way. The process water generally contains some levels of oil/grease, suspended solids, and metals. Table 4-1 presents a number of treatment/disposal options; however, site-specific factors may limit the options available at a particular site. These factors include site location, discharge flowrate, regulatory requirements, contaminant concentrations/loadings, and contaminant types. Appendix B gives further information on water treatment and disposal.

In general, it is more economical to pursue disposal options that do not require some form of pretreatment. However, factors such as site location and availability of off-site wastewater treatment facilities must be considered when determining the best method of disposal. At locations where an on-site WWTP is available, it may be more economical to discharge the process water to these facilities. At remote sites, where off-site treatment facilities are not readily accessible, it may be more cost-effective to obtain

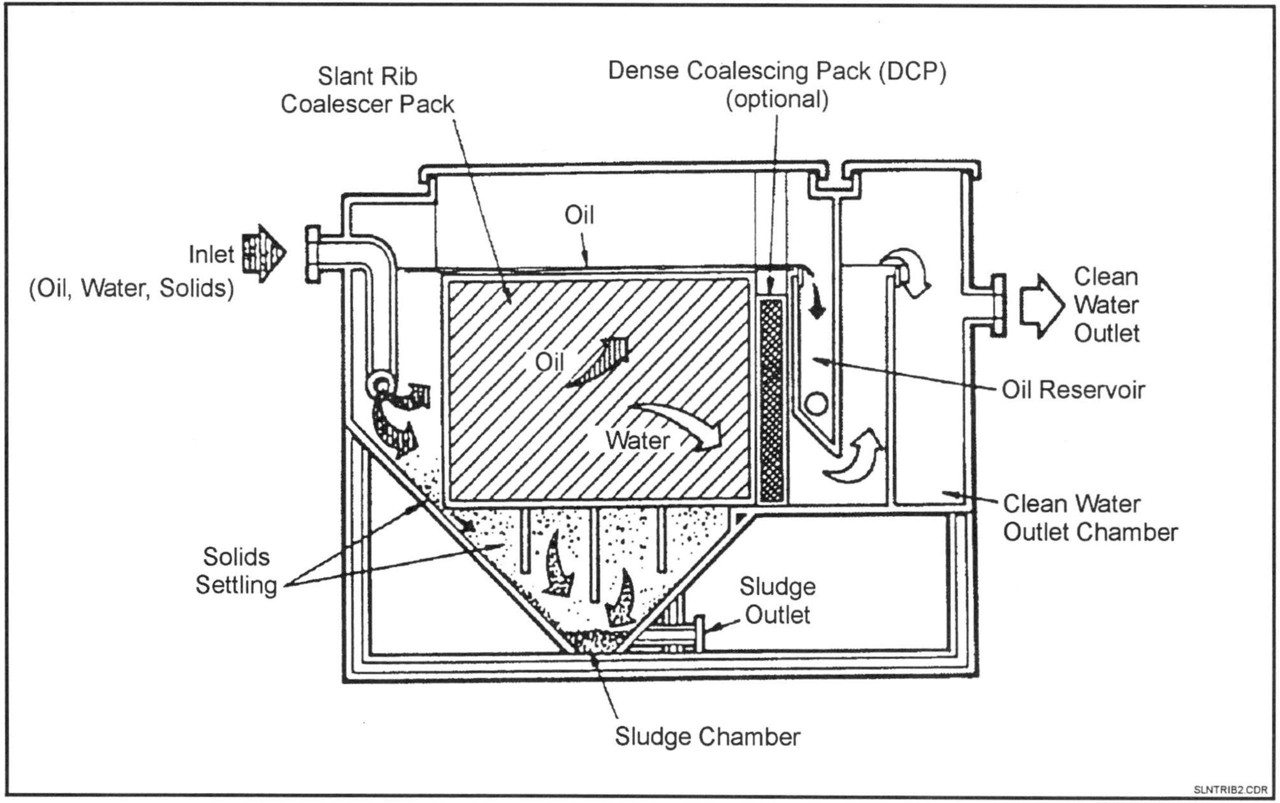

FIGURE 4-7. Oil/water separator (Source: Parkson Corporation Great Lakes Environmental® Products, 1993)

TABLE 4-1. Aqueous treatment and disposal options

Treatment Options	Disposal Options
Chemical treatment including coagulation, flocculation, and precipitation	Transport off site to a disposal facility via truck
Dissolved air flotation (DAF)	Discharge to a storm sewer
Air stripping	Discharge into a WWTP via a sanitary sewer
Carbon adsorption	
Anthracite/clay adsorption	Reinject into the subsurface

a NPDES permit, use some form of pretreatment, and discharge the treated water to the surface water.

The flowrate of the process water from the full-scale bioslurper process must be considered when evaluating disposal options. For example, at one extended bioslurper demonstration site, the water discharge flowrate ranged from 0.1 to 5 gallons per minute (gpm). Discharging the process water to a sanitary sewer with treatment at the WWTP was estimated to cost $0.05 per gallon of water. It would be economical to discharge the process water to the sanitary sewer at 0.1 gpm flowrate but not at 5 gpm. Because the flowrate was expected to be on the high end, on-site treatment was selected.

In addition to the site location and discharge flowrate, other factors such as contaminant types, concentrations, loadings, and requirements by various regulatory authorities would dictate the selection of disposal/treatment options. In general, high concentrations of contaminants posing adverse effects on human health and the environment most likely would trigger pretreatment requirements before discharge.

After disposal options have been considered, negotiations with the appropriate regulatory agencies and/or treatment facilities may be pursued. Regulations imposed by these authorities often affect the cost-effectiveness of the selected disposal/treatment method. For instance, at a site located in northern California, the San Francisco Regional Water Board (SFRWB) required that the discharged water not be allowed to pond on the ground. An automated water distribution system had to be installed to sprinkle the treated water over a large grass-covered area. Although the water treatment system was still cost-effective, the stipulation imposed by the SFRWB resulted in increased capital and operating costs.

At a large number of sites, some form of pretreatment is required prior to discharging the effluent from the OWS. The types and concentrations of the contaminants in the water combined with regulatory requirements usually dictate the selection of appropriate technologies. For example, at a gasoline-contaminated site where the majority of the TPH partitions into the off-gas stream rather than to the aqueous stream, water treatment may not be required if the effluent is discharged directly to a WWTP. However, if the effluent is discharged to the ground surface, some polishing (such as activated carbon adsorption) may be required. If the gasoline contains lead, removal of the lead likely will be required prior to discharge. The most important design criterion is that the treatment technology chosen must reduce the concentrations of contaminants to levels that meet site-specific regulatory requirements. Table 4-2 lists several water treatment technologies that have been used to treat OWS effluent.

TABLE 4-2. Water treatment technologies

Technology	Applicability	Limitations
Coagulation/ flocculation combined with DAF	Excellent technology for removing high concentrations of emulsified oils. Has been shown to remove greater than 99% of oil/grease with influent concentrations greater than 6,000 ppm Can be designed to precipitate and remove heavy metals	• Higher capital costs • Requires extensive O&M • May require treatability tests to determine chemical dosage
Air stripping	Useful for treating water containing low concentrations of high-volatility dissolved hydrocarbons.	• May require off-gas treatment • Cannot be used to remove low volatility contaminants
Carbon adsorption	Competes with air stripping for treating water containing low concentrations of dissolved hydrocarbons	• May be fouled by free-phase product • May have premature break-through if not properly used • Must disposed of after breakthrough and replaced
Hydrophobic clay/anthracite adsorption	Treats water containing low concentrations of emulsified hydrocarbons (<100 ppm) Does not effectively treat BTEX	• Often has handling problems • May have premature break-through if not handled properly • Media must be disposed of and replaced
Settling tanks	Provides residence time allowing emulsified oils to separate Only useful for separating unstable emulsions and floating solids	• Requires large volume and space • Not useful for stable emulsions • May present handling problems

4.6 Stack-Gas Treatment

Because bioslurper systems are considered by most states as small emission sources, they may not require stack-gas treatment, depending on the emission quantities and the site-specific regulatory requirements. In fact, the states vary widely in their air emission regulations according to a U.S. EPA (1991) survey conducted in 1989. For example, 24 of the 50 states have no statewide air discharge standards, and many of the general emission source laws have been written primarily for large sources such as power plants. However, nine states require permits for small point source systems. Of these nine states, two require permits only for sites that discharge more than 10 tons of volatile organic compounds (VOCs) per year.

Among the states with air discharge standards, 17 states express discharge limits on a mass per time basis, but the allowable limits vary widely. For example, North Carolina allows up to 40 lb per day to be discharged, whereas the District of Columbia allows only 1 lb per day. Other states have compound-specific emission limits. Connecticut, for example, lists more than 100 compounds with allowable limits based on both an 8-hour average and a 30-minute average. Some states require that the vapor concentration in the influent stream be reduced by up to 85%.

Selection of a vapor treatment system depends mainly on contaminant concentrations in the stack-gas stream. Figure 4-8 displays several vapor treatment methods and the respective concentration ranges for which they can be used. For TPH concentrations ranging from 10 to 500 ppmv, granular activated carbon (GAC) adsorption may be an option. For TPH concentrations ranging from 10 to 10,000 ppmv, vapor reinjection or biofiltration may be a preferred option. Vapor reinjection would be a favored treatment approach if the soil permeability is high and the location of subsurface structures does not present a problem. For stack gas with even higher concentrations (i.e., 500 to 100,000 ppmv), incineration or thermal destruction in an internal combustion engine (ICE) would be an option. Each of these treatment options is summarized in Table 4-3 and further described in Appendix C. More detailed process descriptions and cost and performance data have been reported by Vatavuk (1991) and Mukhopadhyay and Moretti (1993).

4.7 Safety Switches and Equipment

To minimize the chance of liquid spills or other undesired discharges of liquid or vapor, the bioslurper system should be equipped with a series

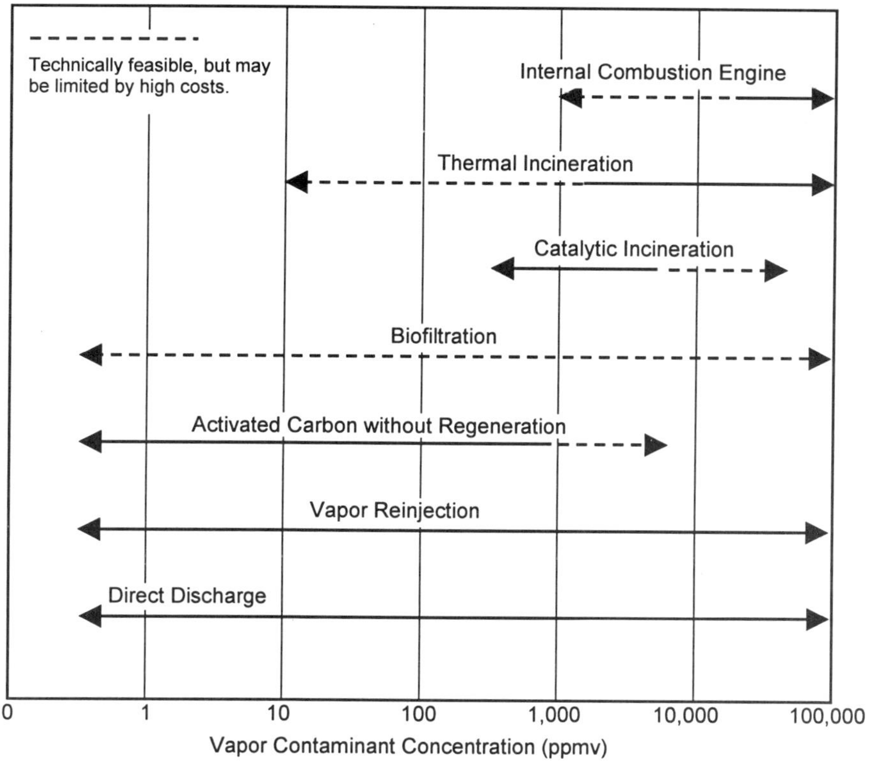

FIGURE 4-8. Concentration ranges for vapor treatment methods

of safety switches. If tripped, either by a high liquid level or a mechanical failure, these switches will open a circuit, causing the LRP to shut down. A safety switch should be placed in any tank that will contain recovered groundwater or LNAPL, and on any water or vapor treatment equipment that will allow the discharge of untreated wastestreams in the event of system failure.

LRPs are equipped with a high-level float switch in the seal water tank to prevent liquid discharge through the exhaust stack. Additional high-level shutoff switches should be installed in the OWS, fuel storage tank, water transfer tank(s), and any further water treatment system. Depending on the design of these tanks, more than one switch per tank may be necessary. If transfer pumps are used to transfer water from tanks to a treatment system, the pump can be wired through a relay that will cut power to the transfer pump if the LRP is shut down. This is done to prevent spills

TABLE 4-3. Stack-gas treatment technologies

Air Treatment Technology	Applicability	Limitations
Granular activated carbon adsorption	Useful for vapor streams containing low TPH concentrations (<100 ppmv). Readily available Relatively low cost if used appropriately. Low O&M requirements.	Limited to use with low TPH concentrations. May require a dehumidfier/demister before treatment vessels.
Biofiltration	May be used for vapor streams with 50 to 5,000 ppmv of TPH	Some time may be required to establish microorganisms in treatment system. Works best with uniform flowrate and contaminant loading.
Reinjection	Useful for a wide contaminant range. Generally can be constructed easily (if the stack-gas volume is relatively small).	Must have relatively high biodegradation rates in the vadose zone. Requires air permeability testing and biodegradation testing. Requires surface emissions testing and may require vapor-phase modeling. May not be permitted near buildings.
Catalytic oxidization	Capable of treating vapor with TPH concentrations in the range of 100 to 10,000 ppmv. Operates at a much lower temperature than flame incinerators resulting in lower fuel costs, reduced NO_x production. Less severe operating conditions for the incinerator. Therefore, longer life expectancy possible compared to flame incinerators.	Requires careful monitoring to prevent overheating of the catalyst. Maximum allowable TPH concentration without dilution is 25% of the lower explosive limit (LEL) for the fuel. Treating vapor with halogenated compounds, sulfur-containing compounds, or nitrogen-containing compounds will deactivate a conventional catalyst. Relatively high capital costs and O&M costs.

TABLE 4-3. Stack-gas treatment technologies (continued)

Air Treatment Technology	Applicability	Limitations
Thermal oxidation flame incineration	Capable of treating vapor with TPH concentrations in the range of 1,000 to 10,000. Most units are easily adapted for catalytic oxidation, if inlet vapor concentration decreases.	Maximum allowable TPH concentration without dilution is 25% of the LEL for the fuel. Somewhat lower capital cost compared to catalytic oxidation.
Flameless incineration	Very efficient at treating stack-gas streams with TPH concentrations up to 10,000 ppmv. Able to handle flowrates between 100 and 1,500 scfm. Minimal energy requirements for vapor streams with TPH concentrations >200 ppmv. Can be used to treat stack gas containing chlorinated solvents.	Higher capital costs than flame incineration or catalytic incineration.
Internal combustion engine (ICE)	Useful for stack-gas streams with concentrations between 14,000 and 300,000 ppmv without supplemental fuel. May be able to perform work when the system is being operated.	Vapor streams with TPH concentrations below 14,000 ppmv require a supplemental fuel to operate the ICE. May require considerable O&M such as regular tuneup of the engine, system adjustment for efficient operation. Difficult to obtain. May be very loud when operating. Cannot be used when halogenated compounds are present in the vapor stream.

in the event that a high level or other problem with the treatment system is the cause of the bioslurper shutdown. In this case, the transfer pump would be pumping water from a tank to a completely full or inoperative water treatment system, causing a spill or discharge of untreated water.

This transfer pump relay system also may be used when pumping from one tank to a second, larger tank. However, a simpler method to prevent spills in this case is to set the high-level switch in the second tank at a level sufficient to contain the volume pumped from the first tank. For example, if the pump in the first tank sends 200 gallons of water to a second, 1,500-gallon tank each operating cycle, the high-level switch in the second tank should be set no higher than the 1,300-gallon mark. This will ensure that the high-level switch in the second tank will cause the entire bioslurper system to shut down. Even if the transfer pump in the first tank is beginning its cycle, the second tank will have capacity sufficient to contain the volume pumped from the first tank without creating a spill.

Mechanical water or vapor treatment systems should be fitted with a relay switch that will cause the bioslurper to shut down in the event of their failure to prevent the discharge of untreated waste. An additional relay may be installed on any treatment system requiring a supplemental fuel supply, such as an ICE or thermal oxidizer. The purpose of this switch is to shut down the treatment unit if the bioslurper shuts down, thus preventing unnecessary consumption of supplemental fuel. All of these safety switches should be wired into the starter box of the LRP, so that the ring pump power supply circuit will be opened if any of the switches are tripped.

An annunciator panel may be installed as an intermediate step between the safety switches and the starter box. The purpose of an annunciator panel is to allow the operator to quickly determine the cause of a system shutdown. The panel consists of a series of relays, lights, and reset switches mounted inside a weatherproof box. If the LRP shuts down as a result of a motor problem, a high- or low-level situation in the seal water tank, a high-level situation in an accessory tank, or mechanical failure of a treatment system, a corresponding light on the annunciator panel will turn off. Each light should correspond to one of the safety switches, and a label should be placed adjacent to the light indicating which safety switch it represents. As soon as the condition that has caused the system shutdown has been identified and corrected, the reset button located beneath the tripped relay can be pressed and the system may then be restarted. An hour meter also should be included in the annunciator panel to provide the operator with the total hours of operation and the approximate time of any system shutdown. A schematic diagram of the electrical configuration of an example annunciator panel is included as Figure 4-9.

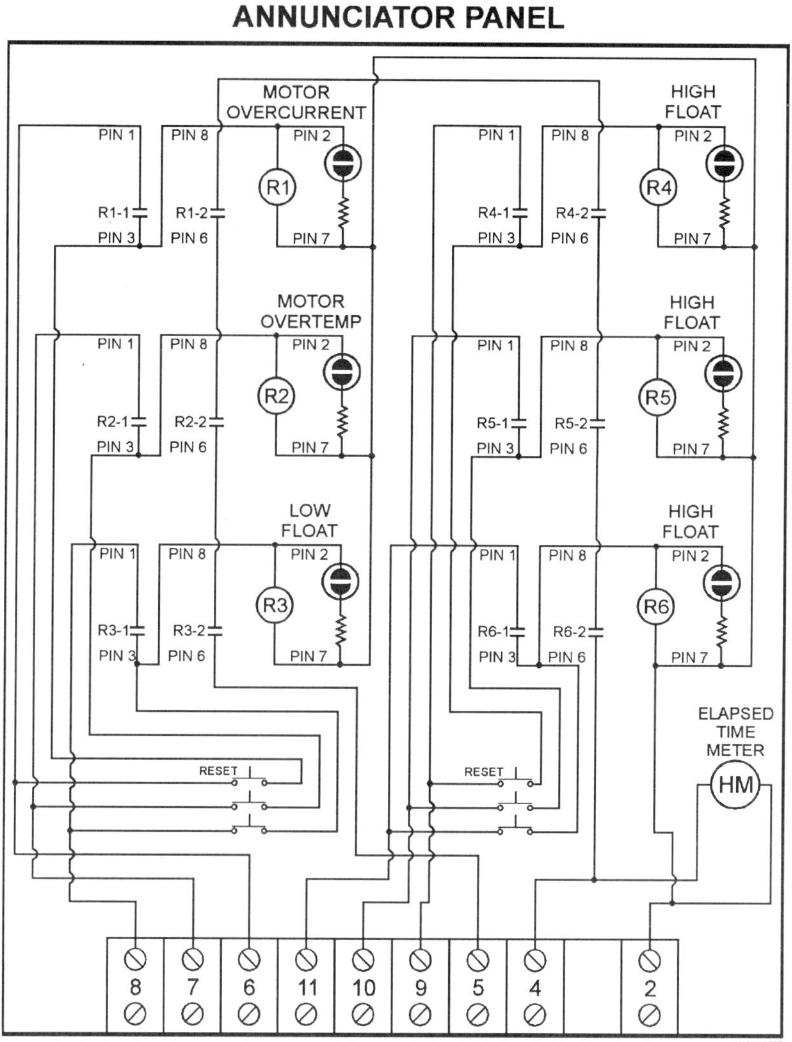

FIGURE 4-9. Electrical configuration of an example annunciator panel

4.8 Life-Cycle and Cost Analysis

The mass removal rate achieved by bioslurping and other LNAPL extraction technologies varies over time. Typically, a period of relatively rapid and steady mass removal rate is followed by a period of exponential decay. This reflects the relative availability of gross contaminant reservoirs compared to less accessible stores of contaminants. As more remote locations of the site are accessed by the extraction mechanisms, the mass removal rate decreases and eventually levels off asymptotically, or approaches zero. Figure 4-10 illustrates the typical relationship between mass removal and time. The duration over which the removal rate remains on the upper plateau depends on the gross quantity of readily available free product and soil characteristics. Some bioslurping systems that have been running for several weeks exhibit LNAPL removal rates similar to those at the time of startup. The overall removal rate at a site is the sum of the removal rates observed in each individual extraction well during operation. Therefore, an operational strategy can be developed to maintain long-term extraction rates near initial conditions by extracting from wells that appear to be producing LNAPL. By eliminating the wells that contain and produce small quantities of LNAPL, the recovery ratio of groundwater to LNAPL decreases as well, thereby reducing the treatment costs for stack gas and process effluent.

The objective of life-cycle cost analysis is to determine the appropriate ratio between capital equipment and O&M commitments to achieve the greatest mass removal in the most cost-effective manner. Capital outlays can be underutilized at sites where the plateau period is brief because the purchased componentry is operated well below its design capacity for the

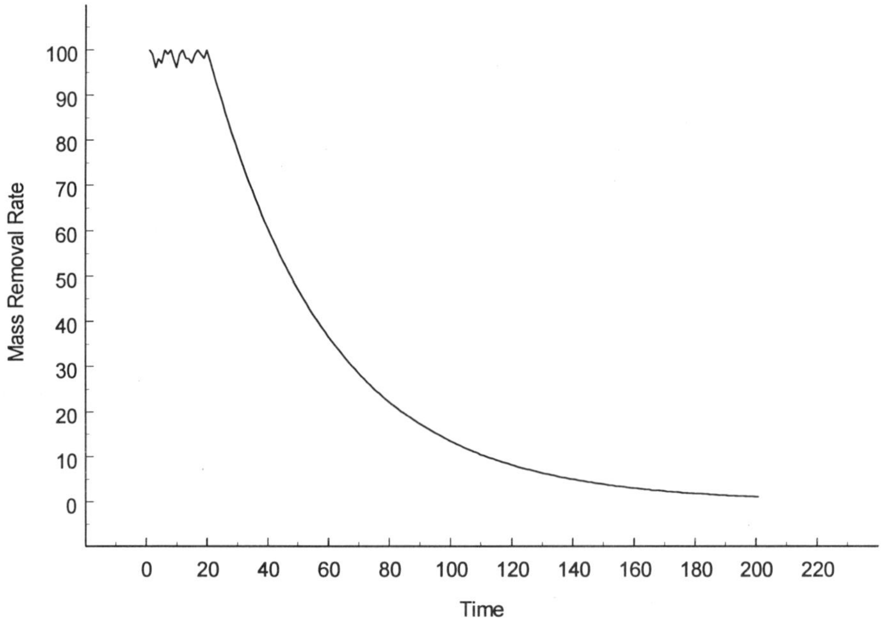

FIGURE 4-10. Typical mass removal curve (rate vs. time)

majority of the operational period. Systems designed to accommodate extraction rates near the maximum may achieve the desired mass removal in a shorter operational period.

A cost-effective approach is to design a system to perform at 40 to 60% of the expected maximum (initial) mass removal rate. With this approach, the lower mass removal rates are sustained near the capacity of the equipment for a longer period of time, making capital outlays more efficient. Another approach for cost-effective treatment of the vapor and process water from the bioslurper involves renting stack-gas and water treatment equipment during the period of high mass removal. When the mass removal rates decrease, the treatment equipment may be returned. This approach also may be used when vapor discharge permits are being obtained. Once the regulated discharge rates have been met and treatment is no longer required, the treatment equipment may be returned.

The initial mass removal rates and the rate of decrease in the mass removal rates may be predicted through interpretation of the results of the mass removal rates during the pilot test. A curve similar to that in Figure 4-10 may be generated using pilot test data. A curve can then be fitted through the pilot-test data to estimate the long-term mass recovery rates. If a pilot test was not performed at the site or if the pilot-test results were inconclusive, the treatment equipment should be designed for half of the initial mass recovery rate during the full-scale effort.

The bioslurper system components that have the greatest incremental capital costs are associated with wastestream storage and treatment/disposal. Incremental costs for LRPs and OWSs are relatively low over the typical range of capacities. Systems designed for greater mass removal rates must be designed to handle greater water and vapor flowrates. Typically, the initial fixed costs for vapor and water treatment are much higher than for the bioslurper unit itself. Furthermore, the incremental costs associated with increasing the treatment unit capacity usually increase at a greater rate than do the incremental costs associated with increasing the bioslurper system capacity. Therefore, the life-cycle cost analysis must include treatment and disposal costs, and the expected type and mass flowrate of contaminants in the vapor and liquid wastestreams may drive the design.

4.9 Design Considerations for System Operation

Other design considerations include noise, power, traffic, groundwater table, and weather factors. These are discussed in Sections 4.9.1 through 4.9.5.

4.9.1 Noise Abatement

Operation of the bioslurper system can result in the generation of noise levels significantly above background levels. Elevated noise levels are caused primarily by the LRP, the gasoline- or diesel-powered generator (for power supply), and/or the ICE (for stack-gas treatment).

One source of noise from the LRP typically is caused by cavitation, which occurs when the pump receives a large volume of water but an insufficient amount of airflow. This problem may be minimized by slightly opening the atmospheric bleed-in valve to allow a slight amount of air to flow into the pump. However, the air that enters the valve also makes a noise. As a result, the preferred solution would be to install a pressure equalization tank between the extraction manifolds and the pump inlet. The equalization tank may be constructed by modifying a 75-gallon air compressor cylinder with inlet and outlet fittings installed on the lid and a short drop tube attached to the outlet fitting. The knockout cylinder described in Section 4.4.1 for oil/water separation also performs adequately to reduce the surges of groundwater that enter the bioslurper system. This design ensures a steady flow of liquid and vapor to the LRP.

Accumulation of solids in the pump head also creates noise. Due to the small gap between the impeller and the pump head, the presence of excessive solids can impede the smooth operation of the pump, causing the pump and motor to labor. The solids may include calcium carbonate scale, iron oxides, and sediment particles. Regular maintenance of the LRP, such as cleaning the Y-strainers, will reduce the amount of solids moving through the pump. If calcium carbonate scale is formed, a weak acid, such as vinegar, may be added to the ring pump influent to dissolve the scales.

Noise due to operation of a generator or an ICE can be decreased by installing a muffler on the unit. Certain ICE systems can be fitted with a noise reduction adapter kit, which consists of a replacement fan that pushes hot air out of the engine compartment. The hot air is forced upward by a bolted-on metal duct. The duct can be equipped with an extension to force the air even higher and away from ear level.

When operated in a noise-sensitive location, such as a residential area, an enclosure may be constructed of concrete blocks or similar materials to divert the noise upward and away from the receptors. Gates to the enclosure should be constructed with metal or vinyl slats through the chain links and, if possible, should be situated to face away from sensitive areas. Walls should be approximately 8 ft high to divert bioslurper operational noise from ground-level receptors.

4.9.2 Power Availability

Prior to installing a bioslurper system, the availability of electrical power must be confirmed. Power is required to operate the LRP, associated transfer pumps, and vapor and process water treatment equipment. The system can be operated using power supplied by a generator, but this is not recommended for long-term operation due to the burden of refueling and servicing the generator and the creation of additional noise.

Use of a three-phase power supply is recommended, as it reduces pump motor wear in liquid extraction applications. Three-phase power usually is available in 208/230-volt or 460-volt currents. Pumps operating on 460 volts will draw fewer amps at full load than pumps running on 208/230 volts, thus allowing the use of a lower amperage breaker. For example, a 10-hp pump operating on 230-volt, three-phase current will require approximately 26 amps at full load, whereas a 10-hp pump operating on 460 volts will require only 13 full-load amps.

Most accessory equipment, such as sump pumps and diaphragm pumps, require 120-volt, single-phase current, and 30-amp service should be sufficient in most cases. The amperage requirements of the equipment should be determined prior to installing circuit breakers.

Because operation of the bioslurper system produces petroleum hydrocarbon vapors that are potentially explosive, the National Electrical Code requires intrinsically safe, explosion-proof electrical installation. All electrical components in the immediate vicinity of the bioslurper system must meet National Electrical Manufacturers Association (NEMA) standards for hazardous locations. NEMA Type 7-rated equipment corresponds to Class I, Group A, B, C, or D ratings of the National Electrical Code, and must be vapor-tight and weatherproof. Conduit joints are finished with seal-off fittings, which are then packed with flame-resistant fiber and cement. A power shutoff switch also should be installed to allow the electricity supply to be blocked in an emergency situation.

4.9.3 Pedestrian and Vehicular Traffic

The presence of unauthorized personnel in the immediate vicinity of the bioslurper system should be avoided when possible due to the presence of liquid fuel, hydrocarbon vapors, elevated noise levels and other hazards during bioslurper operation. To minimize public contact with these hazards, pedestrian and vehicular traffic controls should be instituted. In some cases, bioslurper systems are located in areas of restricted public access, such as in fuel storage facilities. Signs should be posted to alert personnel authorized to work in the area of hazards present at the site, and these workers should be instructed to remain clear of the bioslurper system.

If a bioslurper system must be installed in areas where public traffic is likely, steps must be taken to limit access to the bioslurper and its associated equipment and piping. An enclosure with lockable access gates can be constructed to isolate the bioslurper. Where possible, the well field and manifold system should be contained within the fence or installed subgrade, both to protect the public from tripping hazards and to prevent damage to the system.

In situations where fencing is not practical, wells should be installed with an outer steel protective casing. Each well may be surrounded by three or four concrete-filled steel bollards in lieu of or in addition to the steel protective casing. These provide excellent protection from damage caused by vehicles. Manifold lines should be marked with high-visibility reflective tape and should be placed in trenches if they must extend across a road or sidewalk. In especially sensitive areas, the entire manifold and well system may be installed below grade. This option is considerably more expensive, requiring trenching and the installation of flush-mounted well access boxes.

Barriers should be placed around the bioslurper equipment to protect the system from vehicular traffic. Temporary concrete dividers, similar to those used during highway construction, are very effective. In areas where no vehicular traffic is likely but pedestrian traffic is present, one option is the use of cones and caution tape. This option is economical, but is not particularly durable or attractive.

4.9.4 Fluctuating Groundwater Tables

Some bioslurper sites experience rapid and marked variation in the groundwater table due to tidal influences. These fluctuations may cause complications in the operation of the bioslurper system. Tidal conditions make the issues of pump sizing and drop tube placement more critical. Groundwater levels should be monitored regularly prior to installation of the bioslurper to gather data to determine drop tube placement. Water levels should be measured at both high tide and low tide.

Placement of the drop tube is dependent on the magnitude of the tidal fluctuation and hydraulic conductivity of the formation. If the fluctuation is small, the drop tube may be placed at the deepest measured oil/water interface. In cases where the tidal effect is greater, the bottom of the drop tube will have to be raised somewhat above the low tide oil/water interface. At sites where the hydraulic conductivity is low, the drop tube may be placed closer to the maximum depth of the oil/water interface, whereas the drop tube may have to be placed at the average depth of the oil/water interface, or higher, if the hydraulic conductivity is high. The optimum drop tube level would be where the bioslurper is able to extract both liquid and vapor most of the time. Determining this level depends on frequent observations of flow patterns and adjustments of the drop tube.

The sizing of the LRP also may be influenced by tidal fluctuations. The pump size should be more powerful than the size calculated based on the number of wells. The larger pump provides greater capacity, thus allowing it to more easily overcome the rising water level. The more powerful pump would be able to continue extracting soil vapor and free product, whereas a smaller pump might be able to extract only groundwater. Continuous liquid flow is detrimental to the pump and should be avoided, if possible.

Another method for dealing with tidal fluctuations is to install an in-well float switch that will cause the bioslurper to shut down when the tide is nearing maximum level. This will prevent the extraction of excessive quantities of groundwater and allow for the use of a minimum size pump. The in-well float switch also would automatically restart the pump when the water returns to an acceptable level. The inclusion of the restart switch would require extensive rewiring of the LRP starter motor and the installation of an emergency shutdown switch. A further complication resulting from system shutdown is the possibility of freezing. While the bioslurper is operating, the continual circulation of the process water generally will prevent freezing of water within the system. However, when the system is shut down, water in the seal water tank, LRP, OWS, and any other tanks is susceptible to freezing. This is an important consideration, as tidal influences generally are more pronounced at higher latitudes where freezing temperatures are more likely. Because of the complications involved with wiring and freezing, an adequately sized pump should be used and the drop tubes properly positioned to maintain continuous operation during all tidal conditions.

4.9.5 Cold Weather Operation

Operation of a bioslurper system in a cold climate or during periods of freezing weather requires additional design considerations to protect against a freeze and the resultant damage to the system. If the bioslurper extracts groundwater at a reasonable rate, the system is not likely to freeze due to the continual circulation and movement of water. A very low extraction rate, however, may allow the water to freeze in the piping between the well and the pump, or in the tanks downstream from the seal water tank. A system shutdown also may allow freezing of water in each system component, potentially causing damage.

If an extended period of freezing weather is expected, the bioslurper may be shut down and the process water drained from all system components. This would be an effective way to protect the system. However, in areas where temperature variations produce periods of warm weather interspersed with occasional cold snaps, repeated shutdown and system draining may not be a practical alternative. The most common method of freeze protection is to install heating cables, similar to the devices used on domestic water lines during freezing weather. This alternative allows for continued operation during periods of alternating warm and cold weather. A thermostat senses low temperatures and closes a circuit, allowing the current to flow through the cables. Heat produced by this current is transferred to the pipe or hose to which the cables are affixed. Care should be taken to ensure that the cable and wattage levels are compatible with pipe materials. Additionally, the controller for the heating equipment may need to be contained in an enclosure designed for hazardous environments.

4.10 References

Govier, G.W., and K. Aziz. 1972. *The Flow of Complex Mixtures in Pipes*. Van Nostrand Reinhold Company, New York, NY.

Mukhopadhyay, N., and E.C. Moretti. 1993. *Current and Potential Future Industrial Practices for Reducing and Controlling Volatile Organic Compounds*. Center for Waste Reduction Technologies, American Institute of Chemical Engineers, New York, NY.

Parkson Corporation Great Lakes Environmental® Products. 1993. Manufacturer's literature. 29850 North Skokie Highway, Lake Bluff, IL 60044-1192 USA; www.parkson.com.

U.S. Environmental Protection Agency. 1991. *Soil Vapor Extraction Technology: Reference Handbook*. EPA/540/2-91/003. Office of Research and Development, Risk Reduction Engineering Laboratory, Cincinnati, OH.

Vatavuk, W. 1991. *Estimating Costs of Air Pollution Control*. Lewis Publishers, Chelsea, MI.

5

Bioslurper System Operations and Maintenance

After the bioslurper system has been designed and installed, the system will require operational oversight and possibly maintenance. Based on experience at bioslurper sites, a time requirement of approximately 10 hours/week should be expected for system O&M requirements. However, there is large variation in the time required for each site. Sites that include some form of discharge water or stack-gas treatment typically require more O&M hours. In these situations, the O&M requirements may be as high as 25 to 30 hours/week. If direct discharge of the stack gas and discharge of the process water to a WWTP are permitted, O&M requirements may be minimal (<5 hours/week).

5.1 Troubleshooting and Care of the Liquid Ring Pump

The LRP unit consists of a pump with an attached electric motor, a seal water tank, and an explosion-proof control panel. The LRP creates a vacuum using water as a seal between an impeller and the pump head. Compression and expansion of the gasses and liquids in the pump act to create a strong vacuum, up to 27 inches Hg, which creates the lift to extract LNAPL, groundwater, and soil vapor from the wells. The impeller is turned by an electric motor, which is equipped with switches that will shut the system down if the motor overheats or draws greater than 125% of its full-load amperage. Table 5-1 presents a guide for troubleshooting for the LRP unit.

The seal water tank supplies the water necessary for the creation of the liquid seal and the operation of the pump. Water exits the seal tank from the bottom, flowing through a pipe to the pump head. This pipe often is equipped with a shutoff gate valve and a flow-setter valve. The gate valve,

TABLE 5-1. Troubleshooting guide for the liquid ring pump unit

Problem	Probable Cause	Solution
LRP shuts off	Kill switch is activated	Observe the system to determine which switch has caused the shutdown. A low-level switch is located in the LRP seal water tank. Also, high-level switches are placed in the O/W separator and LNAPL storage tank. Determine which switch has tripped and correct the problem. Restart the LRP.
	Pump kicks out on amperage overload	The reset button must be pressed before restarting the system if amperage overload has caused the pump to kick out. If the system can be started by pressing the start button (without pressing the reset button), then the failure is not amperage-related. An amperage overload condition could occur as a result of an improper amperage setting on the motor starter, an unbalanced power supply, pulling large slugs of water from the manifold, a large backpressure, improper seal water flowrate, and scale buildup in the pump. The following instructions detail the procedure to determine the cause of the overload. ▪ *Amperage Setting*: Turn power off at disconnect. Open the panel on the LRP. The overload protector is located on the motor starter. It should be set at 125% of the current required at maximum load (the current required at maximum load is printed on the motor identification plate). Adjust if necessary. Close panel. Turn on power. ▪ *Unbalanced Power Supply*: Measure the amperage drawn by the motor as it is operating. First, shut off the power at the disconnect. Open the round junction box above the control panel. The wires from the right side of the box are the supply wires. Situate the wires so that they can be reached easily with a clamp-on ammeter. Turn on the power. Turn on the ring pump. Clamp the ammeter onto one of the supply wires. Measure the amperage. If amps are high on one leg, there may be a loose wire connection, or the power supply may be unbalanced. ▪ *Pulling Large Slugs of Water*: Turn on the ring pump. Measure the current. When the ring pump pulls a slug of water, the current will rise. If the current exceeds 125% of the current at maximum load for a period of time, the ring pump will shut off. Install a quiet drum between the wells and the LRP. The drum must be vacuum-tight and heavy-gauge. Cut two openings into the top of the drum. Use bulkhead fittings to make vacuum-tight connections to the inlet and outlet hoses. The outlet bulkhead fitting must be threaded from both ends. Thread a piece of pipe inside the drum to the outlet fitting. The pipe should extend approximately halfway into the drum. Seal the drum. Connect the extraction well hose to the inlet port and connect the ring pump hose to the outlet port. Start the system. Measure the current. An alternative solution is to partially close the ball valve located at the LRP inlet to reduce the water flowrate from the extraction wells. In addition, crack the bleed valve to allow a small amount of dilution air to enter the ring pump. The disadvantage of this solution is that the system vacuum will decrease, which in turn could reduce the number of wells from which the system will be able to efficiently extract.

TABLE 5-1. Troubleshooting guide for the liquid ring pump unit (continued)

Problem	Probable Cause	Solution
LRP shuts off (cont'd)	Pump kicks out on amperage overload (cont'd)	▪ *High Backpressure*: The manufacturer recommends that the backpressure remain below 2 pounds per square inch gauge (psig) when the ring pump is operated at 20-inches Hg vacuum. If fouling occurs in the lines on the discharge side of the pump, excessive backpressure may be placed on the ring pump. Clean out the lines. ▪ *Improper Seal Water Flowrate*: If the extraction flowrate is high, the seal water flowrate may need to be reduced; otherwise, the ring pump may pull too many amps and shut off. Check the flowmeter located in line between the ring pump and the ring pump reservoir. Follow the manufacturer's literature for setting the proper seal water flowrate. ▪ *Scale Buildup*: If scale has fouled the inside of the pump housing, the motor may not start. Attempt to rotate the vanes (located at the rear of the motor housing) with a screwdriver. If they will not move, scale buildup probably is the problem. The pump head can be removed to determine if scale is present. If the pump requires descaling, contact the manufacturer for acid cleaning instructions.
	Thermal overload	The LRP motor is equipped with a thermal overload switch. The contacts will open up if the motor becomes too hot. If a thermal overload condition occurs, wait 10 to 15 minutes and restart the system. The reset button does not need to be pressed prior to restarting the system. If it must be pressed, the malfunction is amperage related. Determine the cause of the overload before restarting the pump. Some possible causes include improper seal water level, faulty thermostat, and faulty motor. Check thermostat leads with an ohmmeter. If the circuit is open, the motor has either overheated or burned out, or the thermostat is bad.
	Pump motor turns off when start button is released	The electrical control circuit is open. This will occur when one of the float switches is opened, the thermostat is opened, or the fuse inside the LRP has blown. See above table sections concerning the kill switch and thermal overload. Turn off the power at the disconnect. Check the fuse. If the fuse is blown, there is likely a short in the 120-volt control circuit. Determine the location of the short.
	Scale buildup in pump head	The pump head may need to be disassembled and cleaned, possibly with a weak acid.
	Transformer has shorted out	Turn off the power at the disconnect. Open control panel. Check voltage across X1 and X2 on transformer secondary. The voltage should be approximately 120 V.
LRP loses vacuum	Scale has built up in the y-strainer at the LRP inlet	Clean strainer.
	System vacuum has decreased	Examine all valves to determine if they are in their proper positions. Measure vacuum at ring pump, along the manifold, and at the wells to determine if there is a leak in the system. Repair leak if present.

which must be open during pump operation, may be closed to allow maintenance on the pump or seal water transfer line without draining the seal tank. The flow-setter valve is used to provide control over the rate of flow of seal water into the pump. A y-strainer is located between the seal tank and the pump. The purpose of the y-strainer is to prevent solid particles suspended in the seal water from entering the pump. The filter in the y-strainer must be cleaned periodically to ensure its effectiveness.

It is critical that the flow of seal water to the pump not be impeded, as the pump must not run dry. The seal water tank is equipped with both a high- and a low-level float switch. If the water level in the seal tank falls below the level of the low-level switch, power to the LRP will be shut off to prevent the pump from running dry. The tank should be refilled with water and the cause of the water depletion should be determined prior to restarting the system. One common cause for seal water depletion is a low groundwater recovery rate, particularly when coupled with a high ambient temperature. Continual recirculation of the seal water without groundwater replenishment heats the seal water and causes evaporation. Seal water levels also may be reduced through leakage in the tank or transfer piping. If water levels exceed the high-level float switch, the power to the ring pump again will be shut off to prevent liquid from escaping from the top of the seal tank. This situation may be caused by excessive liquid extraction rates compared to the flowthrough capacity of the OWS, or by another source of backpressure such as a filter. This problem may be solved by increasing the capacity of the separator or other treatment equipment, or by reducing the liquid extraction rate.

The explosion-proof control box contains the starter, transformer, and associated wiring, connecting the wires from the pump motor to the power supply and control switches. Wires leading to high- and low-level shutoff switches are connected to the motor starter in the control box. Typically, three push-button switches are located on the front cover of the control box. These are the start, stop, and reset buttons. The reset button must be depressed prior to starting the system if the system has shut down due to a high- or low-level float, high-amperage, or high-temperature situation.

The LRP is relatively simple to maintain in working condition. The focus of LRP maintenance is to eliminate solids in the pump head and to continually supply seal water to the pump. These two goals may be accomplished by employing some simple preventative measures.

Noise or rough pump operation may indicate the presence of solids in the pump head. The y-strainers in the seal water supply and inlet lines should be cleaned frequently to ensure favorable operation. The frequency of cleaning will vary depending on site conditions. The drain plug at the bottom of the seal tank may be opened to allow solid particles to flow out, preventing them from entering the ring pump. If excessive quantities of solids are entering the system, this may indicate a problem with the extraction wells, such as a damaged casing or a silted-in well. The well contributing the sediment should be isolated from the system until it can be repaired. Another source of solids is the precipitation of calcium carbonate scale. This problem is more common in environments with alkaline groundwater. The pump head may be removed and cleaned with weak acid, which can be introduced into the system while the bioslurper is operating by letting the pump pull it in through a bypass valve.

Another source of rough and noisy pump operation is cavitation, which occurs when the pump is recovering excessive amounts of groundwater and insufficient vapor volume. This problem can be solved by installing a pressure equalization tank between the wells and the ring pump inlet. Use of this tank ensures a smooth flow of liquid and vapor to the pump. It also will eliminate large surges of water, which causes the pump motor to draw excessive amperage and activate the high amperage shutdown switch (see Section 4.9.4).

The LRP, seal water tank, and associated piping should be protected from freezing, because freezing and expansion of water trapped in the system can cause severe damage to the equipment. The system may be protected from freezing in the event of brief or intermittent shutdowns by using electric heating cable. However, if the bioslurper will be shut down for an extended period during freezing weather, the entire system must be drained (see Section 4.9.5).

The ring pump may shut down for several reasons during normal operation, due to the activation of shutoff switches in the pump motor or high- or low-level switches in the seal water tank or other process tanks. If the pump will not restart, the system must be examined to determine what has caused the shutdown. If the cause is a high-level situation in the seal water tanks, the cause should be remedied before restarting the system. Shutdowns due to high amperage are less likely when using a pressure equalization tank, but may be caused by an irregular power supply or the presence of solids in the pump head. A high-temperature motor shutdown may be caused by solids in the pump head or by misaligned parts, forcing the motor to overexert to turn the impeller.

If the pump operates but produces a poor vacuum, the source of the vacuum loss should be determined. One possible reason for low vacuum is leakage in the manifold and wells. This can be tested by closing off the inlet and atmospheric bypass valves on the LRP. If the vacuum at the pump increases to the desired level when these valves are closed, the most likely cause is a leak in the manifold or at the well seal. The source of the leak should be repaired. If the vacuum reading does not reach approximately

27 inches Hg with the two valves closed, the problem is within the pump system. One possible cause for an inefficient pump is insufficient seal water supply to the pump. The seal water supply line shutoff valve should be fully opened, and the y-strainer should be cleaned. The flow-setter valve also should be opened and cleaned. If these components are clear and operating properly, then another possibility is high seal water temperature, which will decrease the pump volume due to the increased vapor pressure. A low groundwater extraction rate will cause the seal water temperature to increase, as the same water will be continually recycled through the pump. The seal water temperature may be measured by monitoring a thermocouple installed in one of the fittings in the lower portion of the tank. If the seal water temperature is elevated, the addition of cool, fresh water will help to decrease it temporarily. If the water recovery rate does not increase, fresh water should be added periodically. Generally, poor seals at the joints between sections of pipe cause vacuum loss in the extraction manifold. If vacuum loss occurs, all of the joints should be inspected and repaired, if necessary.

Low vacuum also may be caused by wear or damage to the pump. The pump head or impeller may be pitted or worn down, reducing the ability of the pump to operate properly. The cover can be removed so that the condition of the interior parts can be observed. The pump head and impeller can be replaced in the field using basic hand tools.

If the pump produces a vacuum with the inlet and bypass valves closed but fails to create any vapor flow with the bypass valve open, the inlet line may be obstructed. The y-strainer on the pump inlet pipe should be opened and cleaned.

5.2 Operation of Oil/Water Separator

In general, OWSs are relatively simple pieces of equipment that do not require a great deal of operation time or maintenance. However, if floating solids are present in the liquid stream, the OWS may not function properly. The floating solids tend to get backed up on oil-skimming weirs, and a high-level situation may result if the floating solids and LNAPL are held back for a long period of time. Additionally, the floating solids and LNAPL do not flow through the discharge pipes easily and will tend to hold back the fluids in the OWS and cause a high liquid level situation.

If floating solids are present in the liquid stream, shutdown of the bioslurper may be avoided by removing the floating solids or by assisting in the movement of floating solids and LNAPL through the OWS. This may be accomplished by manually skimming the floating solids and LNAPL with a squeegee or a strainer screen. The floating solids also may be passed through the system by raising the water level in the OWS until the floating solids and LNAPL are pushed over the oil-skimming weir. This may be accomplished by closing a valve at the water discharge port of the OWS. The valve should be opened just before water starts to pass over the oil-skimming weir.

The OWS should be cleaned on a regular basis to prevent the coalescing media from becoming clogged. The time interval between cleaning will depend on the mass of floating solids present in the liquid stream. A liquid stream with a high concentration of floating solids will require frequent cleaning. Regular cleaning of the OWS also reduces the TPH concentration in the discharge water. During the routine cleaning, any high-level float switches should be checked to ensure that they are operating properly.

5.3 Troubleshooting and Care of Water Treatment Systems

A variety of water treatment options are available. The appropriate water treatment technology (or combination of treatment technologies) should be selected based on site conditions and regulatory requirements (see Appendix B). This section provides an overview of operation and maintenance requirements for several water treatment/disposal options. These include chemical reaction/flocculation (CRF) and DAF, air stripping, GAC, clay anthracite absorption media, gravity settling tanks, and reinjection.

Because water treatment systems vary widely from site to site, this section is intended to provide only a general overview of O&M activities. Such activities also will be highly specific to individual manufacturers of various pieces of equipment. For this reason, it is important to refer to the manufacturer's literature or contact the vendor for routine maintenance, troubleshooting, or system optimization.

5.3.1 Chemical Treatment and Dissolved Air Flotation

Chemical treatment is effective at removing emulsified oil and grease (O&G) and metals; however, extensive O&M can be associated with such a system. Prior to implementing a water treatment system, a treatability test should be performed on representative water samples to determine the types and dosages of chemicals to be added. Dosages may need to be altered throughout the life of the project if decreased loadings are observed in the process water. Dosages also must be recalculated and reset if changes in the process water flowrate occur. A coagulant is added to the first stage of a two-stage reaction tank. A variety of coagulants are available, some of which may require chemical addition for pH adjustment. If the system is being used for metal removal, pH adjustment also may be required. A polymer is added in the second stage. Chemicals generally are fed through automatic metering pumps that must be set at the optimum flowrates to treat the process water to the required discharge levels. Chemical supplies must be replenished as needed. In general, the polymer will need to be diluted from a concentrated form prior to use.

Following the coagulation and flocculation process, DAF is used to separate the flocs from the process water. This involves pumping microscopic bubbles into the water that attach to the oil-laden flocs and cause them to float. The "float" is then transferred to a larger storage tank for settling; the float in this tank must be removed periodically. Heavy solids that settle to the bottom of the DAF system also must be removed periodically and disposed of with an auger located at the base of the unit. A Toxicity Characteristic Leaching Procedure (TCLP) analysis should be performed on the sludge to determine if the sludge can be disposed of as a nonhazardous waste.

The CRF/DAF fabricated by Great Lakes Environmental® Products consists of three metering pumps, two mixers, a belt skimmer, a centrifugal pump, and two pneumatically operated diaphragm pumps. Each piece of equipment must be maintained and must be functioning properly to meet the desired treatment goals. Details regarding maintenance are presented in the manufacturer's literature for each piece of equipment.

5.3.2 Air Stripping

Air stripping is an effective treatment method for process water that contains low levels of VOCs and minimal emulsified fuel. Because air stripping uses ambient air to strip VOCs from the aqueous into the vapor phase, this technology is more effective at sites contaminated with lighter, more volatile fuels such as gasoline and JP-4 jet fuel. Air strippers consist of various pumps and blowers that must undergo routine maintenance to ensure the proper operation of the system.

Air strippers may exist in the form of towers or the new multistage, low-profile units. Either system is susceptible to fouling resulting from biological activity. In the event that this should occur, the stripper must be cleaned according to the manufacturer's recommendations.

5.3.3 Granular Activated Carbon/Clay Anthracite Adsorption Media

GAC is generally used at sites where the water contains fairly low levels of dissolved contaminants. GAC often is used in conjunction with clay/anthracite media, which can be used to treat bioslurper process water containing low levels of emulsified LNAPL. The vessels of GAC and/or clay/anthracite should be plumbed with at least two in series to prevent breakthrough. Three vessels may be used if breakthrough is expected to occur very quickly. If both GAC and clay/anthracite are used, at least two clay/anthracite vessels should be followed by at least two GAC vessels. Such a treatment system is easy to install, but monitoring is required to ensure optimum performance.

Process water should be passed through a bag filter prior to treatment with GAC or clay/anthracite. This helps prevent fouling that can plug the GAC and clay/anthracite vessels. The mesh size of the bag filter can range from a few to several hundred microns. A greater pressure drop is associated with smaller micron bag filters. The mesh size should be selected based on the size of suspended solids in the aqueous effluent stream. The bag filter must be replaced periodically, with the frequency being dependent on the total suspended solids (TSS) concentration and the process water flowrate. Pressure gauges should be installed on each drum in order to monitor backpressure against the incoming contaminated water stream. An increase in backpressure may indicate blockage within the drum or that the media has been spent. A pump should be installed ahead of the treatment system to avoid starving the pump of fluids, thereby potentially cavitating the pump. Also, it is important to select drums that can withstand the maximum discharge pressure of the pump. The clay will tend to swell over time and generate backpressure within the drum. For this reason, the clay needs to be backflushed periodically. The supplier may provide a schedule for the frequency of backflushing. Backflushing should be performed prior to the drum achieving the maximum discharge pressure of the pump.

Sampling ports should be installed prior to, between, and at the outlet of each vessel. These ports are used to monitor contaminant concentrations in the process water and to determine the treatment efficiency of the GAC or the clay/anthracite. A sample should be collected prior to treatment in order to determine hydrocarbon loadings. These loadings should be compared to sorption curves from the manufacturer to estimate the life of the carbon. Sampling between the two drums should be done to determine when the first drum is spent so that it can be replaced prior to depletion of the second drum. When drums are changed out, the second drum should be moved to the first position and a new drum should be placed in the second position. Sampling at the outlet of the last vessel in series should be performed to verify concentrations at the point of discharge.

5.3.4 Gravity Settling Tanks

Gravity settling tanks can be effective at sites containing a relatively unstable oil/water emulsion in the process water. The tanks provide for residence time that allows the oil/water emulsion to separate from the process water. Bench-scale tests should be performed on process water samples collected during pilot-scale testing to determine a satisfactory amount of residence time required for oil/water separation. During full-scale operation, the groundwater recovery rate should be adjusted to achieve the desired residence time as determined from bench-scale testing.

The treatment system may consist of a single settling tank or of multiple settling tanks in series. The tanks should be equipped with a floating skimmer pump to transfer the LNAPL to the LNAPL storage tank. Because the separated LNAPL accumulates on the surface, the settled water should be pumped from the bottom of the tank.

A variety of configurations exist for multiple settling tanks in series. One option is to actively pump water from the bottom of the first tank into the second tank, and from the bottom of the second tank to the point of discharge. The first tank should be equipped with a high-level float switch that causes the LRP to shut off when activated. The second tank should be equipped with a high-level float switch that trips the pump in the first tank. This prevents continued pumping from the first tank when high-level conditions have been reached in the second tank.

A second option is to transfer water from one tank to another under gravity through a pipe connected to the bottom of each settling tank. This results in the liquid levels in both tanks being equal. A pump is used only to pump water out of the last tank. The tanks must be equipped with a high-level float switch that shuts down the LRP in the event that high liquid levels are reached. The disadvantage of this configuration is that if the integrity of the connecting pipe is compromised, the contents of both tanks will be lost.

Pumps for both fuel and water are associated with settling tank treatment, and these pumps should be maintained regularly as directed by the manufacturer. In addition, the walls of the tank should be checked periodically for structural defects. Valves and connecting pipes also should be inspected routinely for integrity. Treatment with settling tanks is relatively simple; however, precautions must be taken when treating tanks containing large volumes of liquids to prevent spills from occurring.

5.3.5 Reinjection

Another option for water disposal is the reinjection of the OWS effluent into a standard vertical well or an underground infiltration gallery. This option is dependent upon regulatory requirements, characteristics of the effluent water, and hydrogeologic conditions. If the reinjection method is executed at a site, the groundwater recovery rate must not exceed the process water reinjection rate. The reinjection rate also must be slow enough to allow adequate infiltration time so that water does not pond at the surface. Monitoring may be required to ensure that the reinjection of process water is not causing expansion of the plume. Sampling requirements will be dictated by the responsible regulatory body and would likely consist of sample collection along the perimeter of the plume.

5.4 Troubleshooting and Care of Stack-Gas Treatment Systems

Vapor treatment systems are required at many sites to achieve the vapor contaminant discharge limits established by the local regulatory authorities. Several methods of vapor treatment are available, and the proper design should be chosen based on site conditions and regulatory requirements (see Section 4.6). Effluent monitoring should be in accordance with the regulatory requirements.

5.4.1 Reinjection/In Situ Biodegradation

Reinjection of the bioslurper vapor effluent to the vadose zone is a technologically simple method of vapor treatment, with a blower being the only required equipment. A tee is required between the bioslurper stack and the blower to prevent the generation of a vacuum between the bioslurper system and the blower. The blower must be of sufficient size to prevent the release of TPH-contaminated soil gas from the tee between the bioslurper and the blower. An electrical relay is required to turn the bioslurper system off if the blower shuts down. O&M of a reinjection system consists of monitoring soil vapor concentrations to determine biodegradation rates and to ensure that contamination is not spreading. Surface emissions testing must be performed to ensure that contaminants are not escaping to the atmosphere. Pressure gauges should be installed on the blower and reinjection piping to observe any changes in the ability of the soil to accept the vapors. In addition, flowmeters and vapor sampling ports should be installed to measure the contaminant discharge rate to the vadose zone.

5.4.2 Granular Activated Carbon

Use of GAC is a popular treatment method, partially due to its reputation for ease of installation and operation. However, carbon vapor treatment systems require monitoring and maintenance to facilitate proper performance. Pressure gauges should be installed on the carbon drums to allow monitoring of any increase in the backpressure against the incoming contaminated vapor flow. An increase in the backpressure indicates that some form of blockage has occurred in the drum containing the activated carbon. It may be necessary to install a blower between the bioslurper stack to prevent excessive backpressure of the gas in the bioslurper system. If a blower is installed, an electrical relay should be connected between the bioslurper system and the blower to turn the bioslurper system off if the blower shuts down.

Sampling ports should be installed ahead of the carbon treatment area, between carbon drums, and at the outlet of the treatment system. The purpose of these ports is to monitor contaminant concentrations in the vapor stream. An increase in posttreatment concentrations would indicate that the carbon is spent and must be replaced. The sampling port between the two carbon containers is critical for observing concentrations, because it allows the operator to determine when the first carbon drum is spent. The carbon should be changed out before the second container becomes depleted.

The effectiveness of GAC units also depends on the temperature and relative humidity of the influent vapor stream. High relative humidity (above 50%) reduces the efficiency of the carbon, as the water vapor will preferentially occupy the carbon adsorption sites, decreasing the ability of the carbon to capture contaminants from the vapor. Methods for reducing relative humidity in the vapor stream include the use of air/water separators, preheaters, and dehumidifiers. Use of these devices adds to the complexity of the relatively simple GAC method of vapor treatment. Liquid-phase water produced during dehumidification must be collected and disposed of, necessitating the use of a storage tank or drum and the means to transfer the water, such as a pump. This water should be disposed of in the same stream as the bioslurper liquid effluent, as it may carry dissolved hydrocarbons.

5.4.3 Catalytic/Thermal Oxidizer

Catalytic oxidizers are somewhat less complicated than pure thermal flame oxidizers, because the presence of the catalyst allows the oxidation reactions to take place at significantly lower temperatures. This reduces fuel consumption and allows the use of less durable materials in the construction of the unit. Depending on the constituents of the stack-gas stream, additional equipment, such as scrubbers, may be required to remove hazardous contaminants from the effluent that are produced during the oxidation of chlorinated solvents. Stack-gas samples should be collected to determine the effectiveness of the unit and the volume of contaminants discharged to the atmosphere. Operators of catalytic thermal oxidation units should follow the manufacturer's recommendations to achieve optimum performance.

5.4.4 Internal Combustion Engine

ICE treatment of bioslurper vapor effluent streams has been practiced for several years. ICEs are popular at sites where hydrocarbon concentrations are very high, because the ICE has been shown to have a greater destruction capacity than other technologies. Although the ICE may be quite effective at treating highly contaminated stack-gas streams, the equipment requires a high degree of supervision and can be maintenance-intensive. Startup of an ICE system requires particular attention. The position of inlet valves on the ICE and the bioslurper system must be adjusted to allow proper combustion to occur. ICEs are equipped with sensors and a computer to manage the ratio of stack gas to supplemental fuel that the ICE takes, but changes in vapor concentrations may be so rapid that the computer cannot keep pace. In this case, the operator must listen to the engine and monitor the gauges on the ICE to determine what adjustments are required and manipulate the system controls accordingly. Improper or delayed adjustments will cause the ICE to stall and shut down.

When the ICE is connected to a bioslurper system, the ICE inlet usually is attached to a tee fitting installed on the bioslurper vapor stack from the seal water tank. The top of the stack is open to the atmosphere, allowing the ICE to pull ambient air down the stack to dilute the bioslurper vapor effluent. The ambient air can bring the oxygen concentration up to proper levels, and can decrease the TPH concentrations to levels the ICE can handle, if necessary. The system should be monitored to ensure that no vapor is discharged from the top of the stack. A pitot tube or other device should be installed to provide data showing that flow in the stack is from the top down toward the tee leading to the ICE. If vapors are escaping to the atmosphere, the bioslurper should be shut down until the cause of the problem can be determined and corrected.

The ICE must be started from its supplemental fuel supply and must be allowed to run until the ICE exhaust temperature reaches 600 to 700°F. This allows the ICE to reach the proper operating conditions for complete combustion before starting the bioslurper, thus preventing the discharge of untreated bioslurper vapors. The supplemental fuel may be either propane or natural gas, although propane is more common for logistical reasons. Liquid propane is recommended, but gaseous propane may be used.

Once the ICE and the bioslurper are running in tandem, less manual supervision is required. The ICE computer is able to handle most changes in well-gas concentration that occur as the systems operate for longer periods. However, the system still requires some supervision to ensure that the propane is not being depleted and that the ICE is running properly. Some units are equipped with a modem, allowing remote dial-up monitoring of the operational parameters recorded by the on-board computer. The data can be printed on site for ICEs that are fitted with a printer. Samples of ICE exhaust should be collected for analysis to ensure that the ICE is reducing the vapor contaminants to the required levels. The local regulatory body should approve sample collection intervals.

Because the ICE is mechanically complex relative to other treatment methods, O&M can be time-consuming and costly. The ICE is essentially a modified automobile engine, and as such it requires regular service, intermittent repairs, and eventually an engine overhaul. The engine is also subject to parts failures and breakdowns, as would any engine operating continuously. ICE operators should follow the recommended maintenance intervals provided by the manufacturer. In some cases, manufacturers' representatives are available to perform the less-routine maintenance tasks and major repairs, when required. The operator or a technician with engine maintenance skills may perform minor maintenance, such as oil and filter changes.

5.5 Health and Safety Issues

Workers may be exposed to a variety of hazards during the installation and operation of the bioslurper system. Hazards may be physical or chemical, depending on the task being performed. Following proper procedures and wearing personal protective equipment (PPE) can help to reduce the health and safety risk to workers, while site control and security can minimize risk to the public.

The installation phase of a bioslurper study includes advancement of soil borings, completion of extraction wells, and construction of the bioslurper system. Installation of wells and soil-gas monitoring points typically requires the use of an auger drilling rig. Proper PPE, such as hard hats, safety shoes, and eye and ear protection, can help to protect workers from the physical hazards involved with drilling operations. Exposure to petroleum hydrocarbon vapors or free-phase fuel also is possible. Care should be taken to minimize exposure, through the use of petroleum-resistant gloves and monitoring of TPH concentrations in the workers' breathing zone. Drilling operations should be halted if TPH concentrations reach and are sustained at levels above regulation limits. Workers should be trained in respirator use, and respirators equipped with the proper filter cartridges should be available should the situation warrant their use.

Construction of the bioslurper system includes, but is not limited to, assembling the extraction manifold and setting up the LRP, OWS, and treatment and discharge equipment. Workers involved in assembling the manifold may be exposed to vapors from the PVC primer and cement used to connect the manifold piping. Care should be taken to avoid prolonged exposure to these fumes. Physical hazards include the possibility of objects striking the head, feet, or eyes. Again, proper PPE can help to reduce these risks. A qualified electrician should perform the wiring required to supply power to the LRP.

Most hazards encountered during the operation of the bioslurper system involve exposure to hydrocarbon vapors or free-phase fuel. These exposures can occur during sample collection or routine system maintenance. Long-cuff petroleum-resistant gloves should be worn when performing maintenance such as cleaning the OWS, or when collecting samples of recovered free product. Tanks containing free product should remain closed to prevent unnecessary fume emissions. Individuals performing maintenance or collecting samples from these tanks should attempt to minimize the amount of time they are exposed to organic vapors.

Exposure limits for chemical substances common to bioslurper sites are listed in Table 5-2. A list of potential hazards and recommended preventative measures is compiled in Table 5-3.

TABLE 5-2. Exposure limits for chemical substances

Compound	Federal OSHA PEL (ppmv)[(a)]	ACGIH TLV (ppmv)[(b)]	Primary Health Hazards
TPH	500	300	Dizziness, drowsiness, irritated eyes
Benzene	1	10	Irritated eyes and nose, headache, nausea, fatigue, carcinogenic
Toluene	200	100	Irritated eyes and nose, nausea, affects liver and central nervous system
Xylenes	100	100	Irritated eyes and nose, nausea, affects liver and central nervous system
JP-5	NA	NA	Irritated eyes and nose, nausea, dizziness, affects liver and central nervous system

(a) OSHA PEL = Occupational Safety and Health Administration Permissible Exposure Limit.
(b) ACGIH TLV = American Congress of Governmental and Industrial Hygienists Threshold Limit Value.
NA = not applicable.

TABLE 5-3. Potential hazards and recommended preventative measures

Potential Hazard	Measure Taken to Avoid Hazard
Flying particulate	Safety glasses should be worn by all site personnel.
Objects striking head	Hard hats should be worn in the vicinity of overhead hazards (e.g., in the drilling rig area).
Objects striking foot	Steel-toed boots must be worn.
Slips, trips, and falls	Attempts must be made to minimize slips, trips, and falls by providing clear footing.
Exposure to organic contaminants	Disposable gloves, coveralls, and boot covers must be worn when sampling contaminated soil and water.
Exposure to free product	Exposure to free product may occur during sampling. Safety goggles, disposable gloves, coveralls, and boot covers should be worn during sampling.
Exposure to organic vapors	Negative-pressure, National Institute for Occupational Safety and Health-approved cartridge respirators should be available to site personnel should conditions warrant.
Electrical shock	All major electrical work (e.g., wiring, control panel construction) should be performed by a qualified electrical contractor. Care should be taken to de-energize and ground electrical equipment prior to any necessary repair work. Before undertaking repair work, the energy source must be either permanently disconnected or temporarily tagged and locked out to prevent the equipment from accidentally energizing. Must meet OSHA 29 Code of Federal Regulations (CFR) 1910.147 Lockout/Tagout Program requirements.
Fire	Open-flame ignition sources (e.g., smoking) must be restricted from the work area. Free-phase petroleum should be stored in appropriate containers. Signs indicating flammable liquids should be posted where appropriate. Appropriate fire extinguishers should be available to site personnel during drilling activities. A fire extinguisher must be permanently located in the site office/lab building.
Noise	The operation of pumps, drills, vehicles, aircraft, and other sources will create areas where excessive noise is present. Field personnel should be able to identify areas with a high noise level. Earplugs/earmuffs should be worn as warranted. Areas with high noise level per OSHA 29 CFR 1910.95 must be marked with appropriate warning signage.
Traffic	Work may be conducted in some areas where there is traffic. Traffic control should be maintained around the job site at all times. Work areas regularly occupied by field personnel should be marked by cones, lights, or barricades. If it is necessary to work in areas with heavy traffic, personnel should wear appropriate reflective clothing.

6 Site Closure

Development of an exit strategy for bioslurper application and an approach for site closure should be an integral part of the design and implementation of bioslurper technology. This requires review and understanding of applicable regulations (e.g., underground storage tank [UST] regulations) and cleanup criteria/standards based on free-product levels in monitoring wells and/or contaminant levels in soil and groundwater. The operator should carefully evaluate how these cleanup goals are going to be applicable to the site of concern. It is important to note that some regulations may require meeting relatively stringent cleanup goals based on TPH levels. In such cases, the property owner should consider proposing risk-based approaches for individual contaminants, such as procedures established by the ASTM on risk-based corrective actions for petroleum-contaminated sites, (ASTM, 1996). Furthermore, some regulatory agencies, such as California Regional Water Quality Board, may not require soil remediation if it can be demonstrated, by conducting leachability tests (i.e., ASTM's Synthetic Precipitation Leaching Procedure tests), that the contaminants do not pose a threat to groundwater.

A generic decision diagram for a bioslurper exit strategy and site closure is given in Figure 6-1. The RPM may modify the strategy to meet any other site-specific cleanup objectives. As presented in the decision diagram, the bioslurper system should be operated under the "standard mode" (i.e., slurper tube at the oil/water interface) until the LNAPL recovery diminishes. It is important to note that this condition may not be achieved simultaneously at all the wells. As such, the system may be operated until free-product recovery is significantly reduced (20% of the original LNAPL recovery rate) or for a few months (e.g., two months), whichever occurs first. At this point, the bioslurper system should be turned off and each well should be monitored over a limited time period (e.g., one week on a daily basis) for the presence of LNAPL. Bioslurping should then be continued only from the wells that contain LNAPL. The system should be turned

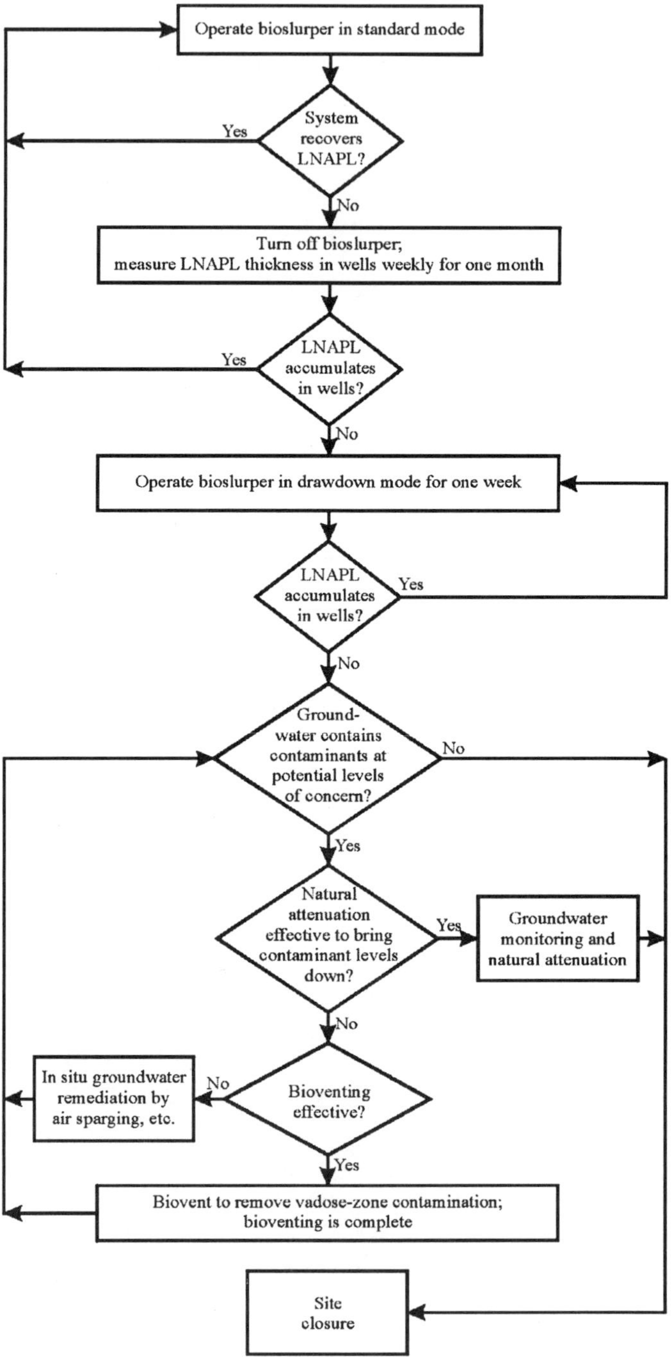

FIGURE 6-1. Generic decision diagram for bioslurper exit strategy and site closure

off again after the LNAPL recovery is significantly reduced or after a few weeks of operation, as discussed above.

At this point, if monitoring shows that LNAPL is detected only in a small number of wells, installation of additional LNAPL recovery wells around those wells should be considered to expedite the remediation activities. When the bioslurper system is restarted to recover LNAPL from these wells under "standard mode," the system may be operated in "drawdown mode" at the wells that do not have LNAPL. The drawdown mode will ensure the removal of any recoverable LNAPL captured in the saturated zone.

When the LNAPL recovery is reduced to 1 gallon or less per week, the RPM should consider shutting down the bioslurper system. A long-term monitoring program should be established at this point to determine the status of the contaminated groundwater plume. If the plume is shrinking or stable, the site may be closed with no further action, or limited long-term monitoring may be performed.

The regulatory agency may require the site owner to demonstrate that natural attenuation of contaminants is taking place. If the plume is not expanding, demonstration of natural attenuation may still be adequate for closure of certain sites. Otherwise, if the leachability studies from selected soil columns indicate that contaminants in vadose zone still contribute to plume expansion, the contractor may elect to operate the bioslurper in "bioventing mode."

Bioventing can effectively remove petroleum-based contaminants from vadose zone (Leeson and Hinchee, 1997). If the plume still appears to be expanding at a rate of potential concern, application of an in situ remedy such as air sparging should be considered.

6.1 References

American Society for Testing and Materials. 1996. *Standard Guide for Risk-Based Corrective Action Applied at Petroleum Release Sites*. ASTM E 1739-95.

ASTM, see American Society for Testing and Materials.

Leeson, A., and R.E. Hinchee. 1997. *Soil Bioventing Principles and Practices*. CRC Press Lewis Publishers, Boca Raton, FL. pp. 244.

A

Bioslurping Implementation Cost-Estimating Guide

A.1 Introduction and Scope

This bioslurping implementation cost-estimating guide is designed to provide reasonably accurate cost estimates for the application of bioslurping technology for the purposes of comparing costs to other feasible technologies. This guide is structured according to the steps that would be followed to test, install, operate, and maintain a bioslurping remediation system. Each section has a descriptive text portion, a table of components and associated costs, and example calculations to estimate the costs of performing the described activity. The following topics are presented in Section A.2:

- Construction of Mobile Pilot Unit (Section A.2.1)
- Pilot-Scale Installation and Testing (Section A.2.2)
- Full-Scale Installation and Testing (Section A.2.3)
- Operations and Maintenance (Section A.2.4).

Section A.3 presents other costs associated with environmental remediation projects. Conclusions are stated in Section A.4.

A.2 Bioslurping Installation, Operations, and Maintenance

A.2.1 Construction of Mobile Pilot Unit

A.2.1.1 Description

Bioslurping field components typically are mounted on a flatbed trailer. Otherwise, site logistics may be such that it is impossible to keep components in close proximity to the affected soils. By utilizing the mobile pilot unit bioslurper system, all field components and materials associated with the bioslurper usually can be kept within 200 ft of the contaminated area. In addition, mobile bioslurper units can be transported quickly and easily from one contaminated site to another. This approach to bioslurping minimizes the costs associated with field mobilization and implementation, and is essential to the overall cost-effectiveness of the bioslurping technology.

A.2.1.2 Table of Components and Costs

Table A-1 presents the items mounted on the trailer that are used to conduct bioslurping at a contaminated location and the cost of each unit that is built onto the trailer.

A.2.1.3 Options

As can be seen in Table A-1, some of the components that are utilized on the bioslurper trailer are optional. The total cost to construct a bioslurper mobile trailer unit without any of the optional components is \$30,175. This is the initial capital cost associated with implementing bioslurper technology. If all the optional system components and discharge water treatment are needed to utilize the bioslurping technology, the total capital cost to construct the mobile pilot unit would be \$34,432. If the ICE is used as the off-gas treatment system, the total capital cost would be \$82,276. If a local power source is not available, it will be necessary to equip the mobile pilot unit with a generator to supply the power required to run the bioslurper components. For short-term pilot testing, it may be cost-effective to rent a generator. If connections to a local power source can be made, the capital cost to build the trailer can be maintained at the costs listed.

A.2.1.4 Calculation

The following equation is the calculation for the total trailer cost to construct the bioslurper system with all the required and optional components and materials:

$$TT = Lr + Mr + Mo_1 + Mo_2 \text{ (or } Mo_3) \quad \text{(A-1)}$$

where: TT = Total costs to construct mobile unit.

Lr = Labor required to construct system trailer = \$9,600

Mr = Materials required to construct system trailer = \$20,575

Mo_1 = Materials optional—Water treatment = \$2,107

Mo_2 = Materials optional—Off-gas treatment = \$2,150

Mo_3 = Materials optional—ICE off-gas treatment = \$49,994.

TABLE A-1. Mobile unit construction components and costs

	Mobile Unit Components	Unit	Cost	Quantity	Total Cost
Lr[(a)]	Labor to construct bioslurper system	hr	$60.00	160	$9,600.00
Mr[(a)]	16-ft trailer with 4-ft ramp, gate, brakes, and spare tire	each	$1,954.00	1	$1,954.00
	10-hp LRP and seal tank	each	$8,516.00	1	$8,516.00
	10 gpm oil/water separator	each	$3,544.00	1	$3,544.00
	500-gal polyethylene storage tank	each	$690.22	1	$690.22
	Digital flow totalizer	each	$222.50	1	$222.50
	Annubar flow sensor	each	$220.00	1	$220.00
	Sump pump	each	$92.86	3	$278.58
	Hand pump	each	$106.65	1	$106.65
	Gasoline hose (¾-inch × 12 ft)	each	$36.00	1	$36.00
	Camlock fittings part, male and female, 1½-inch	set	$16.34	15	$245.10
	Piggyback float switches	each	$30.60	5	$153.00
	2-inch Transporter tank hose	per/ft	$3.67	225	$825.75
	Level control switch	each	$34.35	3	$103.05
	Quick-connect plugs	each	$6.41	10	$64.10
	PVC 2-inch ball valve and tee	each	$45.37	2	$90.74
	PVC 2-inch coupler, schedule 40	each	$3.25	4	$13.00
	PVC 2-inch pipe, schedule 40	per/ft	$0.70	50	$35.07
	Clear PVC pipe	per/ft	$2.04	5	$102.00
	Wellhead seal, 4-inch OD, 2-inch ID	each	$28.40	2	$56.80
	K-type thermocouple plug, male and female	set	$5.56	4	$22.24
	Thermocouple wire (type K), 125 ft	roll	$62.83	1	$62.83
	Pressure gauge 0-30 psi	each	$20.00	1	$20.00
	Vacuum gauge (high) 0-30 inches H_2O	each	$192.85	1	$192.85
	Vacuum gauge (low) 0-10 inches H_2O	each	$192.85	1	$192.85

TABLE A-1. Mobile Unit Construction Components and Costs (continued)

	Mobile Unit Components	Unit	Cost	Quantity	Total Cost
Mr[(a)] (cont'd)	Magnehelic gauge 0-0.25 inch H_2O	each	$54.00	1	$54.00
	Magnehelic gauge 0-0.50 inch H_2O	each	$47.00	1	$47.00
	Magnehelic gauge 0-2.0 inches H_2O	each	$47.00	1	$47.00
	Magnehelic gauge 0-10 inches H_2O	each	$47.00	1	$47.00
	Male connector 68PL-4-2	each	$1.31	10	$13.10
	Male connector 4MSC4N-B	each	$1.52	10	$15.20
	Quick-connect female × ¼-inch tube	each	$12.10	10	$121.00
	Standard brass valve tags 1.5-inch	each	$1.30	1	$1.30
	Nylon tubing ¼-inch (natural)	50-ft pk	$19.25	50	$962.50
	500-gal steel tank	each	$937.00	1	$937.00
Mo_1[(a)]	1,500-gal polyethylene storage tank	each	$825.98	1	$825.98
	55-gal drum activated carbon	each	$496.00	2	$992.00
	PVC check valve, 2-inch	each	$79.14	2	$158.28
	1½-inch transporter tank hose	per/ft	$3.28	30	$98.40
	Camlock fittings part, male and female, 1½-inch	set	$16.34	2	$32.68
Mo_2[(a)]	55-gal drum activated carbon	each	$496.00	4	$1,984.00
	1½-inch transporter tank hose	per/ft	$3.28	30	$98.40
	Camlock fittings part, male and female, 1½-inch	set	$16.34	4	$65.36
Mo_3[(a)]	Internal combustion engine (ICE)	each	$49,994.00	1	$49,994.00

(a) Term is defined in Equation A-1.
ID = inside diameter.

Through the construction of mobile pilot unit phase, the total costs (without options) are:

$$TT = \$\ 30{,}175 \tag{A-2}$$

Any one or all of the optional components can be included or excluded in the total trailer construction costs as required by site-specific requirements. When the ICE is used as the off-gas treatment for bioslurper emissions it can augment the power needed to run the bioslurper system. The capital cost associated with the ICE unit is very high; however, an ICE may be the most viable treatment option for the off-gas emissions. Site conditions and emissions regulations will dictate the type of treatment that should be used as the treatment option.

A.2.2 Pilot-Scale Installation and Testing

A.2.2.1 Description

The overall objective of pilot-scale installation and testing is to determine the feasibility of LNAPL recovery and to properly size the equipment for full-scale bioslurping. The approach of the pilot-scale installation and testing is to initiate pilot-scale bioslurping and to identify the variables that are critical in determining full-scale bioslurping feasibility. These variables include the rate of LNAPL recovery, the ratio of LNAPL to extracted groundwater, the LNAPL ROI (the radius from which free product is mobilized to the bioslurper well), and the efficiency of the system to oxygenate subsurface soils.

If existing groundwater monitoring wells have been found to contain free product, the existing wells may be used for pilot-scale testing of the bioslurping technology. However, if the well construction information on baildown testing data indicate that an existing monitoring well is unusable, it will be necessary to construct a well for pilot testing. Soil-gas monitoring points are used to assess whether bioslurping technology is efficient at remediating the vadose zone soils at the site. If existing soil-gas monitoring points are not present at the site, it will be necessary to install three soil-gas monitoring points in the affected area for the pilot testing. Table A-2 contains the costs associated with installing the necessary groundwater monitoring well and the soil-gas monitoring points for pilot-scale testing.

The field materials required to examine the variables that are critical to successful implementation of a bioslurping operation are presented in Table A-3. These materials will be used again to examine the effectiveness of bioslurping operations during the full-scale implementation of the technology.

TABLE A-2. Pilot-scale test drilling task costs

Pilot Test Drilling Task Description	Unit	Unit Cost
Mobilization to site	Lump sum	$400.00
6¼-inch HAS drill	ft	$12.00
4¼-inch HAS drill	ft	$7.00
4-inch PVC screen	ft	$7.00
4-inch PVC riser	ft	$5.50
4-inch PVC slip cap and plug	set	$25.00
Sand	bag	$9.00
Bentonite	bag	$12.50
Concrete mix	bag	$7.60
Hole plug	bag	$14.00
Well installation	hour	$100.00
Decontamination	hour	$100.00
Steam cleaner rental	day	$75.00
Steel drums	each	$25.00
Per diem	day	$130.00
Cost/ft drilled for monitoring wells	$/ft + $100/well	$29.50/ft
Cost/ft drilled for soil-gas points	$/ft + $100/point	$23.00/ft

Any required permits must be obtained prior to implementing the pilot-scale installation and testing of the bioslurper technology. Typically, a digging permit is required whenever ground will be broken. Additionally, locations may require that both water and air discharge permits be issued to operate a pilot-scale or full-scale bioslurper system. In the State of California, a site investigation permit must be issued before undertaking investigation activities at a contaminant site. Reasonable estimates of the air and water discharged by the bioslurper system can be made from soil-gas and water contaminant measurements prior to test initiation. The analytical results from the pilot-scale testing would be used for the full-scale system discharge estimates.

A.2.2.2 Tables of Components and Costs

Typically, a pilot-scale installation and test will utilize a single LNAPL-extraction well and three soil-gas monitoring points. Table A-2 is a listing of the costs associated with installing a single bioslurper well with a borehole diameter of 2 inches, and three soil-gas monitoring points with 2-inch-diameter boreholes. Depth to contamination and the specific drilling subcontractor used are the variables that affect the cost to install bioslurper and soil-gas wells. Deeper boreholes result in higher incurred costs for drilling. Examples of cost calculations are shown in Section A.2.2.4 of

TABLE A-3. Pilot test materials and analytical costs

	Pilot-Scale Materials/Analytical	Unit	Unit Cost	Quantity	Total Cost
Lr[(a)]	Labor required to conduct the installation and testing	hr	$60.00	320	$19,200
Mr[(a)]	Interconnecting nipple hollow nickel-plated	each	$23.00	3	$69.00
	Jar 250 mL	12/case	$36.50	1	$36.50
	Jar 500 mL	12/case	$44.60	1	$44.60
	Diluter kit OVA purchase	each	$750.00	1	$750.00
	3¼-inch basic soil sampling kit	each	$937.00	1	$937.00
	Brass sleeves, 2-inch × 6-inch	each	$2.10	4	$8.40
	Plastic end caps, 2-inch	each	$0.10	8	$0.80
	Rotary electric hammer and adapter	each	$1,888.00	1	$1,888.00
	Soil probe Handi-jack and adapter	each	$301.15	1	$301.15
	Interconnecting nipple solid stainless steel	each	$18.00	3	$54.00
	Soil-gas probe shaft section, 3 ft	each	$255.00	4	$1,020.00
	Thermocouple readout (Fluke 52)	each	$199.00	1	$199.00
	CO_2 meter	each	$3,700.00	1	$3,700.00
	O_2 meter	each	$1,548.75	1	$1,548.75
	TPH meter	each	$2,075.00	1	$2,075.00
	CO_2/O_2/TPH meter test kit	each	$126.00	2	$252.00
	Oil/water interface probe, 100-ft	each	$1,990.00	1	$1,990.00
	Regulator CGA 590	each	$245.00	1	$245.00
	1/3-hp compressor/vacuum pump	each	$228.00	3	$684.00
	Stopwatch	each	$57.00	1	$57.00
	Erlenmeyer 250 mL plastic flasks	each	$6.27	4	$25.08
	Tedlar® bags	10/box	$82.00	2	$164.00
	Helium gas cylinder	each	$100.00	1	$100.00
	Helium detector	each	$4,500.00	1	$4,500.00

TABLE A-3. Pilot test materials and analytical costs (continued)

	Pilot-Scale Materials/Analytical	Unit	Unit Cost	Quantity	Total Cost
Mr[(a)] (cont'd)	Carbon dioxide, size s3, 10% balance N_2	each	$124.00	1	$124.00
	Hexane, size s3, 4,800 ppm in air	each	$124.00	1	$124.00
	Oxygen, size s3, 10% balance N_2	each	$124.00	1	$124.00
	Latex tubing 3/16-inch ID	100 ft	$45.56	1	$45.56
	Plastic disposable bailers	each	$20.00	2	$40.00
ATr[(a)]	Analysis—TPH and BTEX (soil)	each	$75.00	4	$300.00
	Analysis—Bulk density (soil)	each	$10.00	4	$40.00
	Analysis—Grain size (soil)	each	$50.00	4	$200.00
	Analysis—Particle density (soil)	each	$50.00	4	$200.00
	Analysis—Soil/water (ASTM)	each	$7.00	4	$28.00
	Analysis—Soil/water (UFA)	each	$7.00	4	$28.00
	Analysis—Total porosity	each	$7.00	4	$28.00
	Analysis—TPH and BTEX (water)	each	$75.00	4	$300.00
	Analysis—TPH and BTEX (off-gas)	each	$135.00	3	$405.00
	Analysis—TPH and BTEX (fuel)	each	$75.00	2	$150.00
	Analysis—C-range compounds (fuel)	each	$50.00	2	$100.00

(a) Term is defined in Equation A-3.
OVA = organic vapor analyzer.

this cost estimator. Table A-3 shows the costs for pilot-scale testing materials and analysis.

A.2.2.3 Options

The test wells, field materials, and analytical procedures that are required to conduct proper pilot-scale installation and testing are presented in Chapter 3 of this manual. Site-specific conditions drive decisions when dealing with contaminated sites, and details will vary. In most cases, existing monitoring wells with a known history of free product contamination can and should be used. If no suitable monitoring well or soil-gas points are present, then the drilling costs associated with installing the required wells would be incurred.

The bioslurper well must be established to facilitate the extraction of LNAPL, groundwater, and soil gas from the subsurface to remediate the affected area. The bioslurper well should be located as close to the center of the spill area as possible. The diameter of the bioslurper wells typically is 2 inches. At all sites a 2-inch-diameter bioslurper well will provide the airflow needed to conduct the required test procedures. In addition, 2-inch-diameter wells cost less to install than larger-diameter wells, and frequently extract groundwater at a lower rate, thereby minimizing any water treatment costs that might be incurred during bioslurping. If a larger-diameter well is desired, the cost of installation will exceed that of a 2-inch-diameter well only by the variable rate to use a larger drill auger and the cost of materials.

As is the case with the groundwater, LNAPL may be extracted at a higher rate when a larger-diameter well is employed. Site logistics should ultimately dictate what bioslurper well diameter should be used.

The soil-gas monitoring point boreholes should be 2 inches in diameter. It is not necessary to drill the borehole any larger than the required 2 inches. For sites with shallow contamination and sandy soils, the soil-gas monitoring points can be hand-augured, reducing drilling costs significantly. Additionally, some testing materials listed in Table A-3 are optional (i.e., they are not required for completion of the pilot-scale testing at all sites). The cost of the analytical procedures presented in Table A-3 can be reduced further by limiting the number of tests performed and/or locating a testing facility that has lower costs.

A.2.2.4 Calculation

The following equation is the calculation of costs for the pilot-scale installation and testing of the bioslurper system:

$$PSI = Lr + Mr + ATr + MWi_o + SGPi_o + DC_o \quad \text{(A-3)}$$

where: PSI = Pilot-scale installation and testing costs
Lr = Labor required to conduct installation and testing
Mr = Materials required to conduct installation and testing
ATr = Analytical testing required to conduct installation and testing
MWi_o = Site-specific — Monitoring well installation = [(Depth of MWs) × (number of MWs) × ($/ft drilled)]
$SGPi_o$ = Optional — Additional soil-gas point installation = [(Depth of SGPs) × (number of SGPs) × ($/ft drilled)]
DC_o = Optional — Disposal costs of recovered fuel and extracted water.

Depending on regulatory requirements and on-site disposal facilities, disposal costs (DC_o) are site-specific.

Through pilot-scale installation and testing the total costs are:

$$BIOM_p = TT + PSI \quad \text{(A-4)}$$

where: $BIOM_p$ = total costs through the pilot-scale installation and testing stage.

Exhibit A-1 provides an example problem involving pilot-scale installation and testing of a bioslurper system at an LNAPL-contaminated site and calculates the costs to install, test, and operate the system.

A.2.3 Full-Scale Installation and Testing

A.2.3.1 Description

Using the data interpretation of the pilot-scale installation and testing, the decision about whether to go to a full-scale (multiple bioslurper wells) operation will be made. At sites where LNAPL recovery rates are high, usually greater than 1 gallon per day per well, full-scale bioslurping is a viable LNAPL recovery technology. At contaminated sites where full-scale implementation of bioslurping is utilized, the design, placement, and installation of the multiple bioslurper wells are the most critical factors in achieving successful remediation. Additionally, the approximate depth of contamination will affect the overall full-scale costs. From the data obtained during the pilot-scale operations, the zone of influence exerted by the bioslurper will be calculated. The ROI is the measurement of area where the bioslurper

Example Site X – Pilot-Scale Installation and Testing

The problem involves a fresh diesel fuel spill. The contaminated site is located in an area that has on-site treatment facilities and supplied power sources. It is estimated that the free product comprises about 10,000 gal of fuel over a 2,000-yd^2 area. The work proposal dictates that a short-term bioslurper pilot-scale installation and test be performed at the site to determine the feasibility of full-scale bioslurping to remediate the site. An existing monitoring well, MW-1, has been identified as having a known history of free-product thickness of 2 to 3 ft, and is in the center of the contaminant plume. It is proposed that this well be used to conduct the bioslurper pilot-scale testing. There are no soil-gas monitoring points in the area of concern; therefore, three monitoring points will be constructed during the bioslurper installation. The monitoring points will be placed at distances of 4, 20, and 40 ft from MW-1. The regulatory guidelines for this site stipulate that extracted groundwater must be treated to less than 5 mg/L of total petroleum hydrocarbons (TPH) and less than 1 mg/L of benzene prior to release. Additionally, off-gas emissions are not to exceed 100 lb TPH per day. At contaminated sites comparable to this one, off-gas treatment has not been necessary. However, the extracted groundwater will be carbon-treated to reduce contamination levels to below regulatory requirements.

(Lr)	Labor cost to conduct the short-term pilot test (two workers/320 hr @ $60/hr)	$19,200
(Mr)	Materials cost to conduct the short-term pilot test	$24,807
(GWtr)	Capital cost of water treatment supplies, two 200-lb carbon drums; one 1,500-gal storage tank with accessories	$2,107
(SGPir)	Capital cost to install three soil-gas monitoring points to a depth of 15 ft [(15 ft)*(3 SGPs)*($23/ft)] + [(3 SGPs)*($100/SGP)]	$1,335
(ATr)	Capital cost to conduct the analytical testing on four H_2O, three off-gas, four soil, and two fuel samples	$1,779
(PSI)	**Pilot-scale bioslurper system installation and testing costs**	**$49,228**
(TT)	**Total costs to construct the mobile unit**	**$30,175**
($BIOM_p$)	**Total costs through the pilot-scale installation and testing stage at Example Site X**	**$79,403**

EXHIBIT A-1. Description of pilot-scale bioslurper for Example Site X, with estimated installation, testing, capital, and operating costs

will aerate the contaminated soils and mobilize the free product and groundwater to the extraction well. To implement full-scale bioslurping, enough bioslurper wells need to be installed to ensure that the zones of influence encompass the entire contaminated area.

Figure A-1 depicts a typical layout of an affected area. The ROI is calculated during the pilot testing. The installation of extraction wells is the critical cost factor associated with installing a full-scale bioslurping operation because all materials (i.e., transporter hose, PVC-piping, etc.) used to operate the bioslurper system will be based on the number of wells that are installed. Also, the LRP size (number of horsepower) is determined by the number of wells installed. With an increasing number of wells, the amount of groundwater that might need to be treated also increases. Therefore, in order to minimize the cost to install the full-scale system, the number of wells installed must be minimized to cover only the contaminated area.

A.2.3.2 Table of Components and Costs

Table A-2 presents the costs associated with drilling. As is the case with the pilot-scale installation and testing, the full-scale installation and operation are site-specific. The costs to install the number of wells necessary

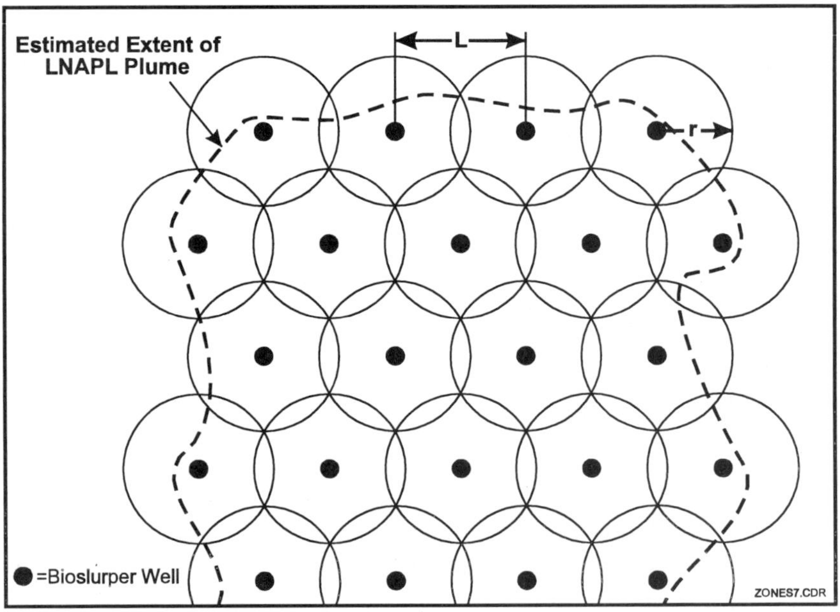

FIGURE A-1. Typical layout of monitoring wells for extended bioslurper testing

to cover the entire contaminated area will vary depending on the site conditions and the volume of the contaminated soils and groundwater. The number of soil-gas monitoring points to be installed for a full-scale bioslurper operation is dependent on the proximity of the wells to one another. If the ROI is high (greater than 50 ft), the number of soil-gas monitoring points can be reduced from three per well to two or one per well. The soil-gas monitoring points should extend radially out from the extraction wells, and at least two soil-gas monitoring points should be located outside the contaminated area to monitor background conditions. Ideally, the number of soil-gas monitoring points will be minimized to reduce installation costs. The monitoring points will be placed in such a manner that they will allow the system operators to gather accurate information about the effectiveness of the bioslurping technology to remediate the contaminated soils.

For the full-scale operation of the bioslurper system, the materials necessary to run the system are shown in Table A-3. The analytical procedures to be used in determining if the bioslurper is remediating the site also are shown in Table A-3. The volume of LNAPL that is being recovered, and the analytical test results will determine if bioslurping is remediating the contaminated site.

A.2.3.3 Options

Thus, the overall costs associated with remediating a contaminated site via bioslurping can be cost-effective compared to costs for other remediation technologies. The site logistics necessary to substantially reduce capital costs are the availability of on-site supplied power sources, wastewater treatment or sewage facilities, and direct release of bioslurper stack gas. If supplied power is available and treatment components are not required, the capital costs associated with installing a bioslurper system decrease drastically. For example, at a remote site where supplied power resources are unavailable, an electrical generator would need to be obtained, adding significantly to full-scale installation costs. By using on-site treatment facilities and power resources, the capital and operating costs can be reduced to make the bioslurper system the most effective and rapid remediation technology for LNAPL-contaminated sites.

A.2.3.4 Calculation

The following is the calculation for the full-scale installation and testing costs (costs will vary from site to site):

$$FSI = Lr + Mr + ATr + MWi_o + SGPi_o + GWt_o + OGt_o \qquad (A\text{-}5)$$

where: FSI = Full-scale installation and testing costs
Lr = Labor required to expand to full-scale installation
Mr = Materials required to expand to full-scale installation
ATr = Analytical testing required to conduct full-scale installation
MWi_o = Site-specific — Monitoring well installation = [(Depth of MWs) × (number of MWs) × ($/ft drilled)]
$SGPi_o$ = Site-specific — Soil-gas point installation = [(Depth of SGPs) × (number of SGPs) × ($/ft drilled)]
GWt_o = Site-specific — Groundwater treatment
OGt_o = Site-specific — Off-gas emissions treatment.

Through full-scale installation and testing the total costs are:

$$BIOM_{fs} = TT + PSI + FSI \qquad \text{(A-6)}$$

where: $BIOM_{fs}$ = total costs through the full-scale installation.

Exhibit A-2 is a continuation of the problem involving the installation of a full-scale bioslurper system at the example petroleum-contaminated site.

A.2.4 Operations and Maintenance

A.2.4.1 Description

Typically, the operations and maintenance required to run the full-scale bioslurper system are minimal. The system requires only weekly on-site presence to collect the data critical in analyzing performance. However, daily visits should be performed to ensure normal operation of the bioslurper. On-site personnel also maintain the system in an operational condition. The personnel will be required to replace and repair any system component malfunctions and/or failures.

The LRP used to establish the vacuum gradient in the bioslurper wells has a service life of approximately 10 years, and should not need to be replaced; however, it may be useful to clean the pump head periodically to maintain the high level of efficiency that the pump needs to create the liquid vacuum. The other components used with the bioslurper system have service lives of three years or less and may need to be replaced before the site remediation is completed.

The primary duties of the on-site personnel will be to record the amount of fuel recovered, groundwater extracted, and off-gas emitted to the

Example Site X — Full-Scale Installation

During the pilot-scale testing, oil/water interface probe measurements in the existing monitoring well indicated that approximately 2.5 ft of floating free product is present in the existing monitoring wells. The depth to groundwater at the site is 12 ft and the soil is a fine sand with a horizontal hydraulic conductivity of 0.5 m/day. Recovery rates from the pilot-scale testing in the monitoring well mentioned were 1.2 gal/day. Therefore, it is recommended that full-scale bioslurping be implemented to remediate the contamination. The radius of influence of the bioslurper was calculated to be 36 ft. The soil analysis indicated that the soils in the unsaturated zone are smeared with free product (i.e., contain TPH and BTEX in 100 kg/L levels). Figure A-2 depicts the contaminated area with nine installed bioslurper wells and the 21 soil-gas monitoring points needed to encompass the entire site. From the water and off-gas emissions data it has been determined that it is not necessary to treat off-gas emissions; however, it will be necessary to treat the water extracted by the system prior to release. Because this is not a remote location, supplied power is available. All the bioslurper wells will be plumbed into a central manifold box, and the vacuum established in each well will be provided by a 10-hp liquid ring pump. The capital costs associated with expanding to a full-scale bioslurper system at this example site are shown below. The cost of the storage tanks and groundwater treatment materials are included in the materials cost.

(Lr)	Labor cost required to expand to a full-scale bioslurper system (2 workers/160 hr total @ $60/hr	$9,600
(Mr)	Materials cost to expand to a full-scale bioslurper system	$15,222
(MWi_r)	Labor and capital cost to install 8 bioslurper wells to a depth of 15 ft [(15 ft)*(8 MWs)*($29.50/ft)] + [(8 MWs)*($100/MW)]	$4,340
($SGPi_r$)	Labor and capital cost to install 18 soil-gas points to a depth of 15 ft [(15 ft)*(18 SGPs)*($23/ft)] + [(18 SGPs)*($100/SGP)]	$8,010
(ATr)	Capital cost to conduct the analytical testing, 12 H_2O, 12 off-gas, 16 soil, and 12 fuel samples	<u>$8,816</u>
(FSI)	**Total labor and capital costs to expand to a full-scale bioslurper system**	**$45,988**
(PSI)	**Pilot-scale bioslurper system installation and testing costs**	**$49,228**
(TT)	**Total costs to construct the mobile unit**	**<u>$30,175</u>**
($BIOM_{fs}$)	**Total costs through the full-scale installation stage at Example Site X**	**$125,391**

EXHIBIT A-2. Converting pilot-scale bioslurper to full-scale for Example Site X, with estimated expansion costs

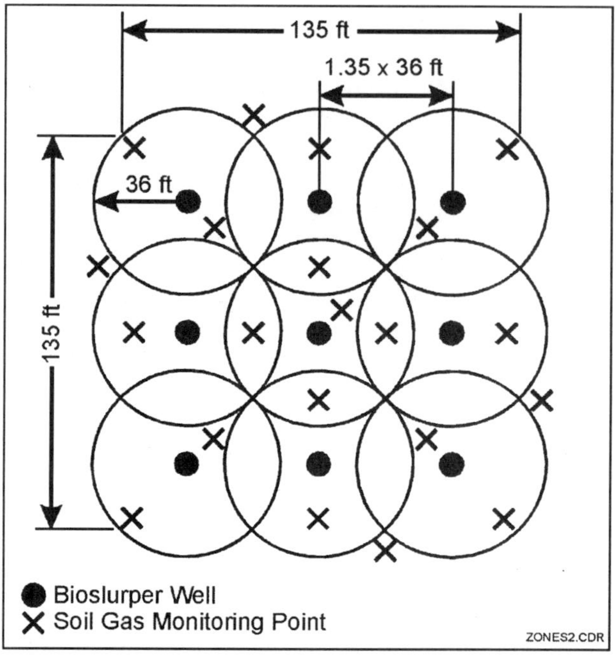

FIGURE A-2. Monitoring well and soil-gas point layout for Example Site X

atmosphere per day. On-site personnel also will be responsible for ensuring that wastewater and off-gas discharges do not exceed any existing regulatory permit allowances.

A.2.4.2 Table of Operations and Maintenance Guidelines

Table A-4 presents the operational actions and maintenance requirements that on-site personnel will perform during bioslurper operations. The daily on-site involvement of site personnel will not be significant, unless there is system component failure.

A.2.4.3 Options

To minimize the costs associated with maintaining the bioslurper system, it is essential that mechanically inclined staff be available on-site to identify, replace, and repair any system component malfunctions. It is also important to use technical staff with field expertise to ensure proper data collection of the critical parameters that define the bioslurper's effectiveness

TABLE A-4. Operational actions and maintenance of the bioslurper system

Frequency	Action
Weekly	Check system components to ensure normal operations Measure and record the fluid levels in each monitoring well Measure soil-gas composition in each soil-gas point Measure water discharge flowrate Measure off-gas discharge flowrate.
Monthly	Sample water discharge and ship for analysis. Sample off-gas discharge and ship for analysis. Adjust the extraction tubes in each monitoring well to the measured oil/water interface
As required	Collect all relevant data to determine if the system is functioning at normal levels (pressures, temperatures) Record fuel recovered during operations and proper removal of the fuel recovered to the disposal or recycling area.

in remediating the site. By separating the functions with two staff members, the labor time required for the bioslurper O&M can be minimized to cost-effective levels.

A.2.4.4 Calculation

The following equation is used to calculate the total O&M costs associated with implementing the bioslurper system for extended testing:

$$OMT = Lr + Mr + KWr + DCr \tag{A-7}$$

where: OMT = Total O&M costs
Lr = Labor required to conduct O&M
Mr = Materials required to conduct O&M
KWr = Power required to run the generator and other components
DCr = Disposal and discharge costs required to remove recovered fuel and extracted H_2O.

Through full-scale O&M the total costs are:

$$BIOM_{om} = TT + PSI + FSI + OMT \tag{A-8}$$

where: $BIOM_{om}$ = total costs through full-scale installation and O&M of the bioslurper system.

O&M costs are site-specific. They will be based on the total costs incurred with repairing and/or replacing materials associated with maintaining the bioslurper system in a normal operational status, as well as the disposal costs to remove recovered fuel and extracted water.

Referring back to the sample problem described in Exhibit A-2, the O&M costs incurred during the past month of operation in remediating this type of contaminated site are presented in Exhibit A-3. This is not a reflection of every site and, depending on site conditions and system component malfunctions, the monthly operating and maintenance costs will vary.

A.3 Other Costs

The cost of implementing a bioslurping system is presented in Section A.2. This section (A.3) includes approximate costs for additional items such as design, work plan preparation, and postremediation/closure sampling. These costs can vary significantly depending on the site conditions and the local, state, and/or federal regulatory requirements. As such, unit costs presented here should be revised for each site.

A.3.1 Design Costs

The costs associated with designing a bioslurper system are site-specific and vary with factors such as formation heterogeneity, natural and factitious obstacles, climate, and local regulatory requirements. Typically, engineering design costs represent approximately 15% to 20% of the cost of installed system components; however, site-specific complications must be considered. For Example Site X, the expected design costs would be around $19,000 using a design factor of 15%. The calculation:

$$\text{Design Costs} = [(\text{TT}) + (\text{PSI}) + (\text{FSI})] \times \text{DF} \qquad \text{(A-9)}$$

where TT, PSI, and FSI have previously been defined and DF is the design cost factor, will yield a reasonable estimate which can be increased further by modifying factors at the user's discretion.

A.3.2 Documentation Costs

Documentation costs also will vary greatly with local regulatory agency requirements. Typically, the cost for the preparation of a Health and Safety Plan is about $10,000. Development of a pilot-scale work plan may cost about $10,000. Preparation of a full-scale Remedial Action work plan

Example Site X - Operations and Maintenance Costs

No instrumentation failures occurred during the first four months of system operation. However, during the fifth month of full-scale operation at site X, one wellhead seal cracked and one water tank sump pump failed. Weekly data measurements and analytical samples have been recorded and shipped. The replacement materials were purchased and installed during the weekly routine checkups of the system. During the fifth month of operation, 550 gal of fuel was recovered, and 402,000 gal of groundwater was extracted. The disposal costs for removing fuel from the site are $2.90/gal. From the results provided by the pilot-scale testing at this site it has been determined that the two carbon canisters used in series to treat extracted water will need to be replaced on a monthly basis. Monthly sampling of the discharge water and off-gas emissions will be made to ensure that regulatory guidelines are not exceeded. The system wastewater is pretreated with the activated carbon and discharged at a cost of $5.00/1,000 gal.

(Lr)	Labor costs required to operate and maintain the bioslurper system (2 workers/32 hr/week @ $60/hr)	$1,920
(Mr)	Material costs required to operate and maintain the bioslurper system	$1,114
(DCr)	Disposal and discharge costs required to remove ten 55 gal drums of fuel and discharge 402,000 gal of treated wastewater	$3,605
(KWr)	Power costs required to operate and maintain the bioslurper system (38,000kWh/month @ $0.10/kWh)	$3,800
(OMm)	**Total monthly operations and maintenance costs at Example Site X**	**$10,439**
(OMT)	**Total yearly operations and maintenance costs at Example Site X**	**$125,268**
(FSI)	**Total labor and capital costs to expand to a full-scale bioslurper system**	**$45,988**
(PSI)	**Pilot-scale bioslurper system installation and testing costs**	**$49,228**
(TT)	**Total costs to construct the pilot-scale mobile unit**	**$30,175**
($BIOM_{om}$)	**Bioslurper installation, operations, and maintenance total costs at Example Site X**	**$250,659**

EXHIBIT A-3. Estimated operations and maintenance costs for Example Site X

for regulatory approval may cost approximately $25,000. If a Quality Assurance Program Plan (QAPP) or a Contractor Quality Control (CQC) Plan is required, it usually can be written for $5,000 to $15,000. Project Final Reports describing the methods, materials, data analysis results and conclusions can be written for $20,000 to $50,000 depending on the scope of the project.

Summarizing, documentation costs can reasonably be expected to vary greatly depending on the local physical and regulatory requirements. For the Example Site X described previously, documentation costs are assumed to be $70,000.

A.3.3 Site Closure (Sampling and Analysis) Costs

After routine free-product recovery rate and soil-gas monitoring results indicate that bioremediation rates and residual contamination concentrations have been minimized, soil samples can be collected and analyzed to demonstrate site cleanup. The number of samples to be collected depends on the site size and heterogeneity as well as local regulatory requirements. It is assumed that one soil sample per 50-ft grid over the site will be adequate to characterize the soil at closure. At Site X in the aforementioned examples, the 2,000-yd^2 (18,000-ft^2) site would require approximately 16 soil samples analyzed for TPH and BTEX for closure characterization. At a cost of $100 for each set of TPH and BTEX analyses (Exhibit A-4), the

Example Site X (Summary) — Other Costs

The total cost for implementing bioslurping (installation, operations, and maintenance) at Site X has been determined to be $250,659. This value can be increased by the other costs described above to yield a total bioslurping project cost:

$BIOM_{om}$	=	$251,000
Design Costs	=	$ 19,000
Documentation Costs	=	$ 70,000
Closure Costs	=	$ 3,000
Contingency Costs	=	$ 25,000
TOTAL	=	**$368,000**

EXHIBIT A-4. Total estimated costs for bioslurper operation at Example Site X

analyses would cost $1,600. Sample collection and shipping costs must be added to this figure. It is assumed that for Site X, $3,000 would cover final soil sample collection and analysis. It must be noted that soil formation heterogeneity and local regulatory requirements can increase closure sampling and analysis costs significantly.

A.3.4 Contingency Costs

It is reasonable to expect that unforeseeable circumstances will arise in any project which may add to the total project cost. Health and safety issues, scope increases, climatic interference, vandalism, regulatory delays, and equipment manufacturing errors are only some examples of factors that can add cost to a well-planned project. To cover these costs, a contingency factor ranging from 10% to 20% typically is used. Again referring to Example Site X, assuming a factor of 10%, contingency costs for installation, operation, and maintenance would be $25,000.

A.3.5 General and Administration, Overhead, and Fee

Each contracting company applies these charges to different categories of project costs. The labor rates in this bioslurping costs estimator are assumed to be fully burdened values already containing these fees. Application of other overhead and administrative inflation factors must be computed according to contractor-specific methods.

A.4 Conclusions

This cost estimator is designed to provide readers with a set of useful guidelines that will enable them to make sound decisions in costing and implementing bioslurper technology at a petroleum-contaminated site. By using the information provided, readers will be able to make an informed decision as to whether or not bioslurping would be a cost-effective and rapid technology to use in remediating the contaminated area. The costs presented in the tables of this document reflect the market price of the materials as of May 1996. The drilling costs given in Table A-2 are estimates made from previous quotations of several drilling contractors. The drilling costs will vary depending on the location of the affected area. Mention of manufacturer and trade names does not constitute endorsement of said product by Battelle or NFESC. Equation A-10 provides the final costing factors associated with designing, documenting, implementing, and closing a full-scale bioslurping project:

$$BPC = TT + PSI + FSI + OMT + DDCC = BIOM_{om} + DDCC \qquad \text{(A-10)}$$

where:

BPC	=	Bioslurping project costs
TT	=	Total costs incurred to construct system trailer with optional materials included
PSI	=	Total costs incurred to conduct pilot-scale installation and testing
FSI	=	Total costs incurred to expand to a full-scale installation
OMT	=	Total costs incurred to operate and maintain the system in a normal operational mode
$BIOM_{om}$	=	total costs through full-scale installation and O&M cost of the bioslurper system
DDCC	=	Design, documentation, site closure, and contingency costs.

B

Water Treatment Technologies

B.1 Chemical Treatment and Dissolved Air Flotation

Chemicals may be added to the OWS effluent to remove contaminants such as emulsified O&G and metals. Chemical treatment is very effective at removing these contaminants; greater than 99% removal can be achieved. However, operation and maintenance costs tend to be higher than those associated with other treatment technologies such as air stripping or carbon adsorption.

Chemical treatment usually involves adding inorganic and/or polymeric coagulants at electrical pH values to form flocs, and removing emulsified oil droplets and inorganic contaminants (such as heavy metals) through adsorption, coprecipitation, and other physical/chemical mechanisms. The treatment may be performed in a CRF system, which consists of a two-stage chemical reaction tank, polymer mixing, chemical metering pumps, constant- and variable-speed mixers, and associated instruments and controllers (Figure B-1). The effluent enters the two-stage mixing tank, where a coagulant (such as ferric sulfate) and sodium hydroxide are added separately to the first stage of the tank and a polymer is added to the second stage of the tank. The polymer causes the oil-laden precipitants present in the process water to floc together.

After the chemical treatment, physical separation must be performed to separate the flocs from the process water. DAF is commonly employed (Figure B-2). A DAF system consists of a flotation chamber (including a belt skimmer and a float storage sump), an air compressor, an air-dissolving tank, and associated controls and meters. Microscopic bubbles pumped into the water attach to the oil-laden flocs, causing them to rise to the surface.

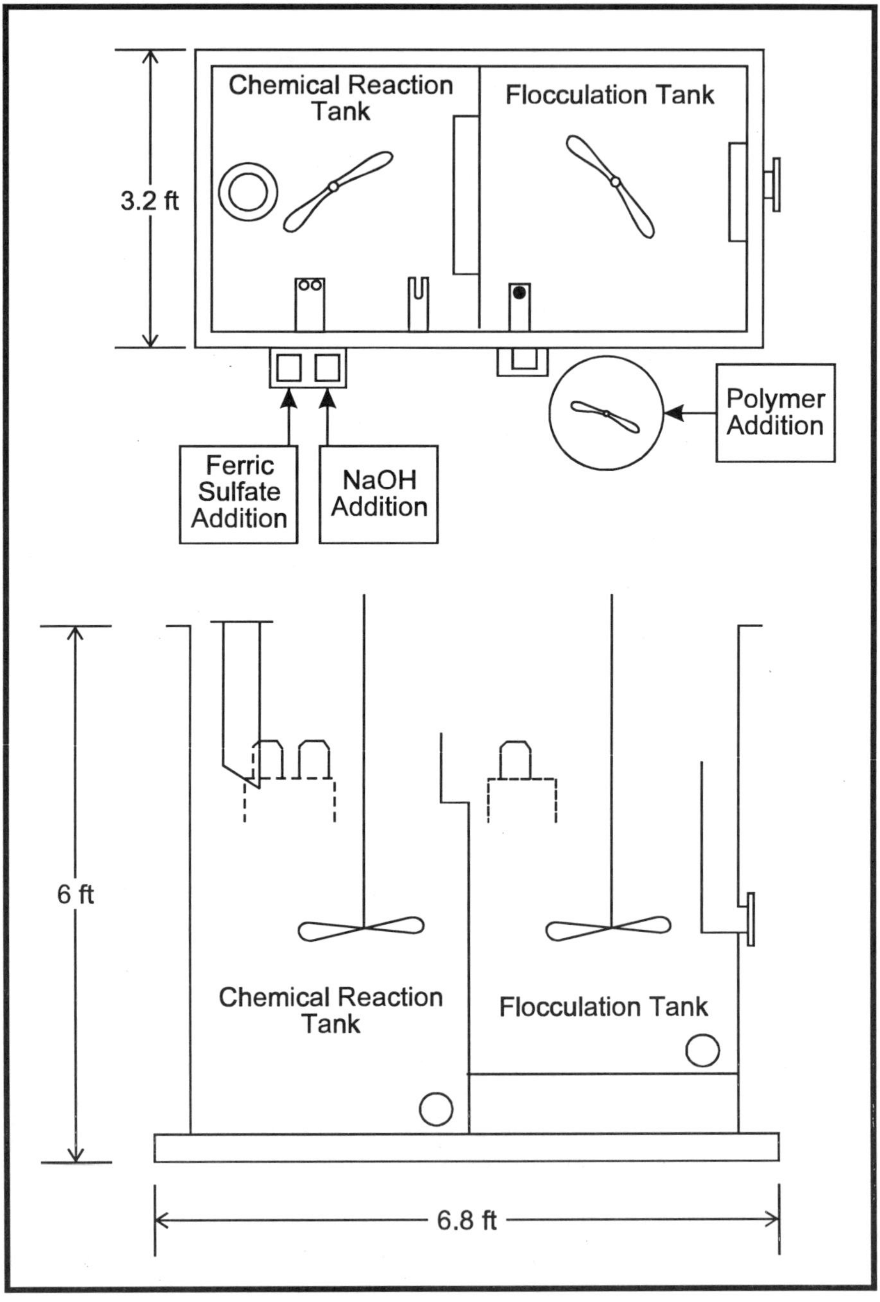

FIGURE B-1. Chemical reaction/flocculation system (Source: Parkson Corporation Great Lakes Environmental® Products, 1992)

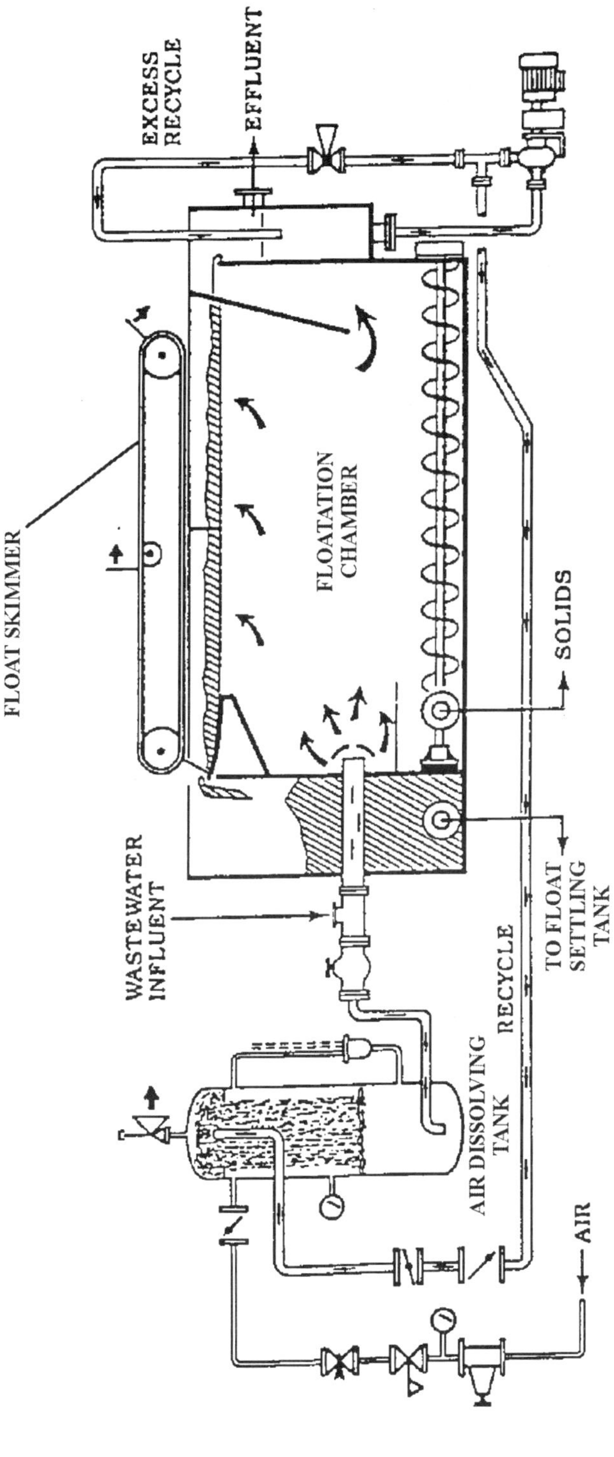

FIGURE B-2. Dissolved air flotation system (Source: Parkson Corporation Great Lakes Environmental® Products, 1991)

The "float" is skimmed by a belt skimmer into a storage compartment mounted inside the unit and then transferred to a large storage tank for additional settling. Heavy solids that settle to the bottom of the DAF system must be removed periodically and disposed of with an auger located at the base of the unit.

Chemical treatment combined with DAF is very effective in removing emulsified O&G from the process water. Table B-1 presents removal results for a treatment system that is being used at a site in northern Florida.

TABLE B-1. CRF/DAF treatment results for TPH

	TPH as Diesel	
Concentration in OWS Effluent (ppm)	Concentration in DAF Effluent (ppm)	Removal (%)
5,000	45	99.1
27,000	31	99.9
5,000	91	98.2
6,800	8.2	99.9
300	48	84.0
920	5.5	99.4

Although the CRF/DAF treatment system is designed primarily to remove O&G, it can be used to treat heavy metals by raising the pH values of the process water to precipitate the metals of concern. Figure B-3 shows the optimum pH for removing various metals. At the site located in northern Florida, the pH of the process water was maintained at about 9 to remove copper, lead, and zinc. The removal results for lead and zinc are presented in Table B-2.

The CRF/DAF process generates a waste sludge, which is composed primarily of metal, oil, and water. The following site-specific factors will affect the rate of sludge production:

- Concentration of emulsified oil in the process water
- Concentration of TSS in the water
- Process water flowrate
- Process water pH
- Coagulant addition rate.

A TCLP analysis should be performed on the sludge to confirm that the sludge can be disposed of as a nonhazardous waste. Analyses for percent

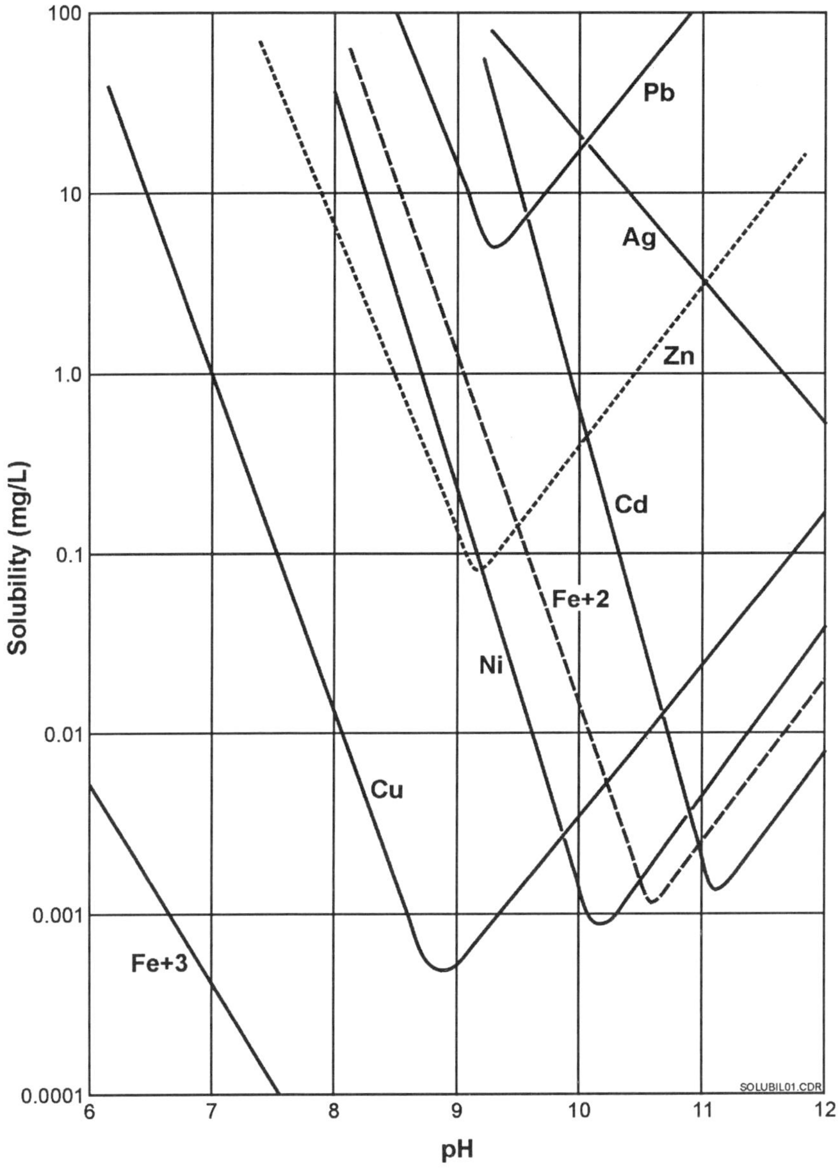

FIGURE B-3. Optimum pH for removal of various metals

O&G, moisture content, and TSS also should be performed. At some sites, such as the site in northern Florida, the percent O&G was high enough (about 40%) to allow the sludge to be recycled for its heating value; hence, the disposal charge was minimal.

TABLE B-2. CRF/DAF treatment results for metals

Lead			Zinc		
Concentration in OWS Effluent (ppm)	Concentration in DAF Effluent (ppm)	Removal (%)	Concentration in OWS Effluent (ppm)	Concentration in DAF Effluent (ppm)	Removal (%)
1,430	3.3	99.8	6,210	123	98.0
120	2.1	98.2	697	<20.0	>97.1
62	<20	>67.7	1,070	<200	>81.3
154	<20	>87.0	1,820	<200	>89.0
30	<20	>33.3	396	<200	>49.5
61.7	<2.0	>96.8	25.1	<20.0	>20.0

O&M requirements for the CRF/DAF are significantly greater than for other treatment technologies such as air stripping and carbon adsorption. The equipment contains a number of valves, pumps, and motors. Each component must be maintained properly to meet the desired treatment goals. Also, the operator must calculate and set the dosage rates for the treatment chemicals. If changes in the process water flowrate occur, dosages must be recalculated and reset. The chemical supplies must be replaced periodically. At the site in northern Florida, one drum of ferric sulfate lasts for approximately one month, and one drum of sodium hydroxide lasts about two months. The polymer is shipped in concentrated form and must be diluted for use. A 5-gal bucket lasts about four months. A solution of polymer must be made up every three days because the dilute polymer expires after 72 hours, resulting in poor flocculation.

The CRF/DAF treatment system is used only when emulsified oils and/or metals cannot be treated by other physical treatment processes. Higher O&M costs make this technology less attractive when contaminant concentrations are low. As a rule of thumb, chemical treatment should be considered when TPH concentrations are greater than the allowed discharge limits and cannot be reduced using conventional gravity OWSs. Prior to full-scale implementation, a treatability study using a series of jar tests must be performed to determine coagulant types, coagulant dosages, sludge production volume, and sludge leaching characteristics. The data obtained from these tests then can be used to estimate full-scale capital and operating costs.

B.2 Air Stripping

Air stripping may be used to remove VOCs from the OWS effluent. The process uses ambient air to strip VOCs from the aqueous phase into the vapor phase. Air stripping is useful for treating the process water at sites when the water discharged from the treatment process contains only low levels of VOCs. Air stripping is not very effective in removing emulsified oil from the process water. In fact, emulsified oil may foul the packing media so that additional maintenance is required to clean the stripper.

Site-specific conditions and regulatory considerations will determine if air stripping is applicable at a particular site. Typically, it is applicable at sites that are contaminated with gasoline, avgas, and JP-4 jet fuel. These fuels and their constituents are volatile with high Henry's law constants, making them ideal for stripping. In addition, little emulsion is formed in the bioslurper process water generated at sites contaminated with these fuels. For opposite reasons (i.e., lower Henry's law constants and oil emulsion), stripping will be difficult to implement at sites contaminated with diesel fuel, JP-5 jet fuel, heating oil, and waste oils.

Several different types of air stripping devices are available. Historically, large towers constructed out of metals (such as aluminum and steel) have been used in air stripping applications. The diameter and height of the tower are determined from design calculations that take into account the process water flowrate, the Henry's law constants for the contaminants of concern, and the air flowrate required to achieve the necessary operation efficiency. The tower diameter may range from less than a foot to several feet, and the height may range from 10 to 50 ft. Air stripping towers are constructed with a plate-and-frame design or with a packing material such as polypropylene pall rings or ceramic saddles. The packed towers have been used more commonly in environmental remediation applications. The disadvantages of both types of towers are that they have relatively high capital and installation costs and that they cannot be demobilized easily.

Recently, a new design for air strippers has emerged. These are high-efficiency, multistage, low-profile air strippers that are compact and skid-mounted and can be mobilized and demobilized easily for environmental applications. These strippers use plates or trays instead of packing materials. Multiple stages can be placed in series to obtain high removal efficiencies.

A cost analysis should be performed prior to implementing air stripping for treating the bioslurper process water. The effluent air emitted from the stripper may require treatment. Vapor treatment using gas-phase activated carbon, thermal/catalytic oxidation, or an ICE may be needed to treat this gas. These technologies are discussed in Appendix C.

B.3 Granular Activated Carbon

Similar to air stripping, GAC may be used to treat OWS effluent at sites where the water contains fairly low levels of contaminants. A cost-analysis must be made to determine if it is more cost-effective to use GAC treatment than air stripping. In general, liquid-phase GAC adsorbs dissolved hydrocarbons up to a maximum of 5% of its weight (i.e., a 200-lb drum will adsorb about 10 lb of hydrocarbons). Typically, it is more cost-effective to use activated carbon when the mass loading of hydrocarbons is less than approximately 20 lb/day. However, site-specific factors, such as the process water flowrate, contaminant concentrations, and the expected treatment duration will affect the cost to implement different technologies.

GAC treatment is implemented by installing, typically, two or more 200-lb to 5,000-lb drums or canisters of GAC in series. The process water is pumped from a surge tank through a bag filter housing to the GAC vessels. The bag filter housing is used to remove suspended solids that, if not removed, eventually will plug the GAC vessels. A 200-μm mesh bag filter is generally sufficient to prevent fouling of the GAC. The bag filter must be replaced periodically. The frequency of replacement is dependent upon TSS concentrations and process water flowrates.

The effluent from the first GAC vessel should be monitored regularly for hydrocarbon breakthrough. When the hydrocarbon concentration in the effluent water reaches a predetermined level, the GAC has been exhausted and the vessel must be replaced. The most common practice is to put a new vessel on line after the "old" second vessel. The effluent from the second vessel also should be monitored to ensure that the water quality is in compliance with regulatory discharge limits.

B.4 Clay/Anthracite Adsorption Media

Organo-clay and anthracite can be used to remove hydrocarbons from an aqueous stream. Organo-clay is quaternary amine-modified bentonite clay, which often is mixed with anthracite at a 30:70 ratio. The mixed media may be used to treat the bioslurper process stream containing low levels of emulsified oils. The loading capacities of the mixed media may vary significantly, depending on the contaminant types, free-phase oil vs. emulsified oil, contaminant concentrations, and TSS concentrations. At some sites, secondary treatment with either GAC or air stripping also may be required prior to discharging the water. Table B-3 shows the percent removal of hydrocarbons observed at three bioslurper demonstration sites using the clay/anthracite media.

The clay/anthracite mix may be placed in a number of different types and sizes of vessels. It is important to use vessels that can withstand the maximum discharge pressure created by the pump supplying the process water. As the clay/anthracite mix adsorbs the oil present in the process

TABLE B-3. Percent removal of hydrocarbons using clay/anthracite mix

Site	Average Influent Concentration (ppm)	Average Effluent Concentration (ppm)	Average Removal (%)
Coastal Systems Station (CSS) Panama City	240	55	77
Robins AFB UST70/72	29	20	31
Robins AFB Site SS010	36	22	38

water, the organo-clay will swell, pressurizing the vessel until the maximum discharge pressure of the pump is reached. As with activated carbon, multiple vessels may be placed in series and effluent samples should be collected and analyzed regularly to ensure compliance with regulatory requirements.

One difficulty associated with the clay/anthracite treatment is channeling of the process water in the vessel. As the organo-clay swells, the water to be filtered may channel through one area of the vessel, resulting in reduced treatment efficiency. One remedy is to periodically backflush the vessel with clean water. This, undoubtedly, would create O&M requirements and, perhaps, material-handling problems. For full-scale, long-term treatment, it is recommended that an automated backflushing system be installed to backflush the vessel when the pressure gradient across the vessel reaches a preset value. For short-term or pilot-scale applications, the media may be mixed using a shovel.

B.5 Gravity Settling Tanks

Settling tanks are used to provide residence time to the process water, allowing unstable oil/water emulsions to separate. If the emulsions are chemically stable, a prolonged residence time may not significantly decrease the hydrocarbon concentration in water. During pilot testing, samples should be collected and bench-scale tests should be performed to determine the amount of residence time required to improve the oil/water separation.

The data collected from bench-scale testing and the expected full-scale process water flowrate are used to size the settling tank. If they are needed, and if space allows, multiple tanks can be placed in series to provide additional residence time. The tanks are connected near the bottom to allow the same liquid levels in all tanks. This configuration also avoids the need to pump from one settling tank to another. At many sites, additional treatment may still be required after settling, which would increase the overall treatment costs.

Because the separated LNAPL floats and accumulates on the water surface, the settled water should be pumped near the bottom of the last tank in series. Also, the tanks should be equipped with a floating skimmer pump to transfer the LNAPL to the LNAPL storage tank. Table B-4 shows the percent removal of hydrocarbons observed at various bioslurper demonstration sites using settling tanks.

TABLE B-4. Percent removal of hydrocarbons using settling tanks

Site	Average Influent Concentration (ppm)	Average Effluent Concentration (ppm)	Average Removal (%)
Eaker AFB Site 2	6.5	3.6	55
Dover AFB SS27/XYZ	960	175	82
Hill AFB	180	7.4	96
McGuire AFB	47	38	19
CSS Panama City	1,500	240	84
Robins Site AFB SS010	43	36	36

Note that settling tanks usually occupy a very large floor space. At sites with limited space, it may not be feasible to use settling tanks for the process water treatment.

B.6 Reinjection

A some sites, it may be possible to reinject the OWS effluent into a standard vertical well or an underground infiltration gallery. Some regulatory requirements may have to be met before this option can be considered. These requirements may include contaminant types, contaminant concentrations, TSS concentrations, and other hydrogeologic considerations.

B.7 References

Parkson Corporation Great Lakes Environmental® Products. 1991. Manufacturer's literature. 29850 N. Skokie Highway, Lake Bluff, IL 60044-1192 USA; www.parkson.com.

Parkson Corporation Great Lakes Environmental® Products. 1992. Manufacturer's literature. 29850 N. Skokie Highway, Lake Bluff, IL 60044-1192 USA; www.parkson.com.

C

Stack-Gas Treatment Technologies

C.1 GAC Adsorption

GAC is the most commonly used vapor-phase treatment method. GAC adsorbents possess high surface areas (i.e., 1,000 to 1,400 m^2/g), thus allowing effective contaminant removal. Many vendors sell or lease prefabricated, skid-mounted units that can be put into service with only a few days of notice. The GAC adsorption, however, is economical only for low concentrations. When the vapor concentrations are high, GAC adsorptive capacity may be quickly exhausted. Replacement and disposal of spent GAC can become expensive. The cost of disposal of the spent carbon will be particularly high if the spent carbon is identified as a Resource Conservation and Recovery Act (RCRA)-listed or RCRA-characteristic waste.

The GAC adsorption capacity depends on factors such as influent vapor temperature and relative humidity and, most importantly, the influent VOC types and concentrations. Isotherms, which show the mass of contaminants adsorbed per unit mass of carbon, are available to predict the contaminant-specific adsorption capacity for a specific type of carbon. GAC generally has a high affinity for VOCs, such as hydrocarbons and chlorinated compounds. Some hydrocarbons such as isopentane do not adsorb well.

High relative humidity in the incoming vapor stream limits the effectiveness and increases the cost of vapor treatment with GAC. Water vapor preferentially occupies adsorption sites, thereby decreasing the capacity of the GAC to remove contaminants from the incoming stream. Vendors typically recommend that the relative humidity of the off-gas stream be below 50% prior to entering the GAC adsorbent. Entrained water should be removed by an air/water separator (demister) or by preheating the vapor stream. However, preheating may reduce the effective capacity of the GAC, so a trade-off is involved in selecting the preheating temperature.

As a rule of thumb, the GAC's adsorptive capacity for most hydrocarbons in the vapor stream is about 1 lb of hydrocarbons per 10 lb of GAC. Because the GAC cost including material, canister, O&M, and disposal is about $3/lb, the cost of the GAC treatment would be about $30/lb of hydrocarbons.

Specialized resin adsorbents have been developed and are now entering commercial application for treatment of organic vapors in off-gas streams. These synthetic resin adsorbents have a high tolerance to water vapor. Air streams with relative humidities greater than 90% can be processed with little reduction on the adsorption efficiency for organic contaminants. The resin adsorbents are amenable to regeneration on site. Skid-mounted modules are available consisting of two resin adsorbent beds. The design allows one bed to be on line treating stack gas while the other bed is being regenerated. During the desorption cycle, all of the organic contaminants trapped on the resin are removed, condensed, and transferred to a storage tank. The desorption process used to regenerate the resin is carried out under vacuum using a minimum volume of nitrogen purge gas. A heat exchanger in the bed heats the resin during regeneration. The same heat exchanger is used to cool the bed to increase sorption capacity while it is on line treating stack gas (Downey et al., 1994).

C.2 Biofiltration

Treating stack gas with biofiltration is an innovative technology. Connolley et al. (1995) have used a biological filter to treat both stack gas and water from a pilot bioslurper system. Several other tests have been reported in the literature, indicating effective treatment of influent with 50 to 5,000 ppmv of TPH (U.S. EPA, 1994). For a vapor concentration of 1,000 ppmv, the residence time required ranges from 15 to 90 seconds (Skladany et al., 1994). Insufficient data are available to allow a detailed cost and performance comparison of biofiltration with the more conventional stack-gas treatment.

C.3 Reinjection/In Situ Biodegradation of Vapor Emissions

In situ biodegradation of the bioslurper vapor emissions can be a cost-effective and environmentally acceptable treatment option. Reinjection of the bioslurper stack gas probably will require approval from the regulatory agency. Stack-gas reinjection offers the advantages of low surface emissions and no point source generation. The reinjection treatment consists of distributing the stack gas back into the soil and destroying the contaminants through in situ aerobic biodegradation. Reinjection is accomplished by piping the stack gas to air distribution wells or trenches where the vapor stream infiltrates back into the soil. In situ respiration and soil-gas permeability tests must be performed to determine the biodegradation rate and the ROI required for the reinjection point. The soil volume available must be sufficient to accept the vapor stream and allow biodegradation of the contaminants in the stack gas.

Reinjection wells should be located and designed to ensure that the reinjection process destroys the contaminants rather than increases the contaminant migration. After reinjection is initiated, surface emissions testing must be performed to ensure that contaminants are not escaping to the site surface. Soil-gas monitoring should be performed to ensure that contaminant migration is not being increased. Monitoring of migration is particularly important at sites where air extraction is necessary due to the presence of buildings.

C.4 Destruction in a Catalytic Oxidizer

Catalytic incineration is a thermal treatment process that uses a catalyst to increase the oxidation rate of organic contaminants in an off-gas stream, allowing acceptable destruction efficiency at a lower temperature than for flame incineration. In catalytic incineration, the stack gas is heated and passed through a combustion unit where the gas stream contacts the catalyst. The catalyst accelerates the chemical reaction without undergoing a chemical change itself. The catalyst increases the oxidation reaction rate by adsorbing the contaminant molecules on the catalyst surface. Sorption phenomena on the catalyst serve to increase the local concentration of organic contaminants at the catalyst surface and, for some organic contaminants, reduce the activation energy of the oxidation reaction. Increased concentration and reduced activation energy increase the rate of oxidation of the organics (Kiang, 1988).

The active catalytic material typically is a precious metal (e.g., palladium or platinum) that provides the surface conditions needed to facilitate the transformation of the contaminant molecules into carbon dioxide and water. The catalyst metal is supported on a lower cost, high-surface-area metallic or ceramic support medium.

The metal catalyst and support are exposed to the heated stack gas in a catalytic incineration unit. The catalytic incineration unit uses either a fixed-bed or a fluidized-bed system. Fixed-bed systems include metallic mesh, wire, or ribbon or ceramic honeycomb supporting the catalyst metal or a packed bed of catalyst-impregnated pellets. Fluidized beds also use catalyst-impregnated ceramic pellets but operate at sufficiently high flow to move and mix the pellets during treatment (Kiang, 1988).

The main advantage of catalytic incineration versus flame incineration is the much lower temperature required with a catalyst. Catalytic systems typically operate at 600 to 900°F (CSM Systems, 1989), compared to 1,200 to 1,600°F for flame incineration. The lower temperatures for catalytic incineration result in lower fuel costs, less severe service conditions for the incinerator construction materials, and reduced NO_x production. Natural gas or propane is a typical fuel used for supplemental heating when the contaminated vapor streams do not contain sufficient heat value for self-sustaining incineration. Energy costs can be reduced further by reclaiming heat from the exhaust gases, i.e., using the exhaust gas flow to preheat the influent vapor stream.

Catalytic incineration units require careful monitoring to prevent overheating of the catalyst. Overheating can damage the catalyst metal surface and/or the support, reducing catalytic activity. The allowed influent organic vapor concentration depends on the heat value and LEL of the influent vapor stream. Concentrations exceeding about 3,500 ppm of VOCs normally are diluted with air to prevent excessive energy release rates and to control the temperature in the catalytic unit. Safety is a concern with these units, as with any incineration method. The maximum permissible total hydrocarbon concentration depends on the local fire codes and permitting requirements but is below 25% of the LEL at essentially all sites. The total hydrocarbon concentration in the vapor is continuously measured at the inlet to the catalytic unit to control the dilution airflow during operation.

Treating stack gas containing halogenated compounds, sulfur-containing compounds, or nitrogen-containing compounds will deactivate a conventional catalyst due to chemical reaction of the catalyst metal with halogens or strong sorption of SO_x and NO_x on the catalyst. Catalysts specially designed for treatment of chlorinated compounds by catalytic incineration are available on the market but are more expensive than catalysts for treating petroleum hydrocarbons. The incineration unit typically will require scrubbing to remove acid gases formed when treating halogenated compounds (Buck and Hauck, 1992). The significant cost elements of a catalytic incinerator are the capital cost (or rental) of the unit, O&M and monitoring cost, and the cost of makeup fuel.

C.5 Thermal Oxidation Flame Incineration

Flame incineration converts hydrocarbon compounds to carbon dioxide and water by direct thermal oxidation. Complete destruction of contaminants requires high temperatures, typically 1,200 to 1,600°F, and/or long residence times. In a thermal incineration system the stack gas is mixed and introduced into a refractory-lined combustion chamber where one or more burners supply heat to thermally oxidize organic contaminants. When the influent vapor concentration is low, makeup fuel will need to be added to maintain the temperature required to ensure adequate mineralization. Natural gas or propane typically serves as the supplemental fuel. When the influent vapor concentration is high, dilution air may be needed. For safety reasons, influent concentrations normally are limited to 25% of the LEL (U.S. EPA, 1986). The LEL for gasoline is between 12,000 and 15,000 ppmv, depending on the grade of gasoline (Little, 1987).

Direct incineration usually is inappropriate for influent vapor streams containing chlorinated compounds. Complete combustion of these compounds will generate corrosive hydrochloric acid vapors. Partial or incomplete combustion of chlorinated compounds could result in the production of chlorinated products of incomplete combustion.

The capital cost of a flame incinerator typically is less than that of a catalytic incinerator and, at higher hydrocarbon concentrations, flame incineration may be less costly than catalytic incineration. The flame incinerator operates at a higher temperature than a catalytic incinerator. The supplemental fuel cost for the flame incinerator increases more rapidly than the fuel cost for the other incineration methods when the vapor concentration decreases. Thermal incineration becomes cost competitive when the inlet vapor concentration approaches 25% of the LEL.

C.6 Flameless Incineration

The flameless incineration process converts hydrocarbon compounds to carbon dioxide and water by passing a stack-gas stream through a heated ceramic matrix. The matrix geometry and uniform high temperature of the matrix are reported to give good destruction efficiency for organic vapors in air, without using an open flame. The vendor indicates that this technology has several desirable characteristics for treatment of vapors in stack gas from remediation systems. The removal efficiency is reported to be high and stable over varying operating conditions. Tests have shown efficiencies greater than 99.99%, and this removal is attained continuously (Rubin, 1995).

The matrix is raised initially to the operating temperature of 1,800°F by electrically powered radiant heating. No additional energy input is required if the heat value of the vapors is sufficient. This point is near a concentration of 200 ppmv. If the concentration is below this value, natural gas or propane can be bled in with influent to maintain the proper temperature.

As with any incineration technique, excess air is added to dilute the concentration to safe levels if the influent is too rich. Incineration in a ceramic bed has handled influent vapor concentrations at the 10% level. Throughput levels depend on the model selected (presently ranging from 100 to 1,500 scfm), with higher flowrates met by combining two or more of the modular units.

Flameless incineration units have higher capital cost than flame incineration or catalytic incineration units, but have much lower energy use. The flameless units are most cost-competitive when the off-gas flowrate is high and the contaminant concentration is low. According to the vendor, the greatest advantage of this technique is its ability to mineralize chlorinated compounds without producing chlorinated products of incomplete combustion or degrading the ceramic beads. Mineral acid vapors still would be produced.

C.7 Destruction in an Internal Combustion Engine

ICEs can be used to destroy organic contaminants through oxidation in a conventional engine. ICEs have been used to destroy landfill gas for years. Application of ICEs to destroy hydrocarbon vapors is more recent. The first operational unit was installed in 1986.

The ICE used is an ordinary industrial or automotive engine with its carburetor modified to accept vapors rather than a liquid fuel. The airflow capacity of the ICE is determined by the cubic inch displacement of the engine, the engine speed, and the engine vacuum, and can be estimated by the following equation:

$$\text{capacity} = (\text{rpm}/2)*(\text{CID}/1{,}728)*(0.85)*[1-(\text{EV}/\text{P})] \qquad \text{(C-1)}$$

where: rpm = engine speed in revolutions per minute
CID = engine displacement in cubic inches
EV = vacuum in the engine intake manifold in inches of mercury
P = local air pressure in inches of mercury

A 140-cubic-inch-displacement, 4-cylinder engine running at 2,250 rpm and 10 inches Hg engine vacuum with an atmospheric pressure of 30 inches Hg would have an off-gas treatment capacity of 52 scfm. ICE treatment units are available in sizes from 140 to 920 cubic inches. Currently available ICE treatment units operate the engine near idle conditions. The off-gas capacity may be increased by applying a load to the engine to increase engine speed and decrease engine vacuum (increase absolute pressure in the manifold). Engine loading by attaching a generator to supply power to the site has been proposed but is not routinely practiced.

A second required modification to the ICEs is the addition of a supplemental fuel input valve when the intake hydrocarbon concentration is too low to sustain engine operation. Propane is used almost universally, although one vendor uses natural gas and reports a 50 to 75% reduction in energy cost. The ICEs also are equipped with a valve to bleed in ambient air to maintain the required oxygen concentration. Soil vapor may have very low concentrations of oxygen, especially during the initial stages of operation. Ambient air is added to the engine via an intake valve at a ratio sufficient to bring the oxygen content up to the stoichiometric requirement for combustion.

A diagram of an ICE system manufactured by Remediation Service International (RSI) is presented in Figure C-1. The engine is mated to an on-board computer system, which monitors normal engine operational information and system performance parameters. Data monitored by the computer may be reported using an attached printer. The engine and computer system are contained in a metal housing, and the unit is commonly trailer-mounted, making it easily portable. Safety features include sensors that will shut down the ICE in the event of engine overspeed, high coolant or oil temperature, low oil pressure, fire, or high water in the intake filter. The ICE can be wired to the bioslurper by a relay system, which would cause the bioslurper to shut down in the event of an ICE failure, preventing the discharge of untreated vapors. Relays may also be installed to stop the ICE if the bioslurper shuts down, preventing the unnecessary consumption of supplemental fuel by the ICE.

A catalytic converter is an integral component of the system, providing an important polishing step to reach the low discharge levels required by many regulatory agencies. A standard automobile catalytic converter with a platinum-based catalyst normally is used. Data from the South Coast Air Quality Management District, the air quality regulatory body for Los Angeles and the surrounding area, show that the catalyst reduced TPH concentrations from 478 ppmv to 89 ppmv and from 1,250 ppm to 39 ppm, resulting in important additional contaminant removal (U.S. EPA, 1991).

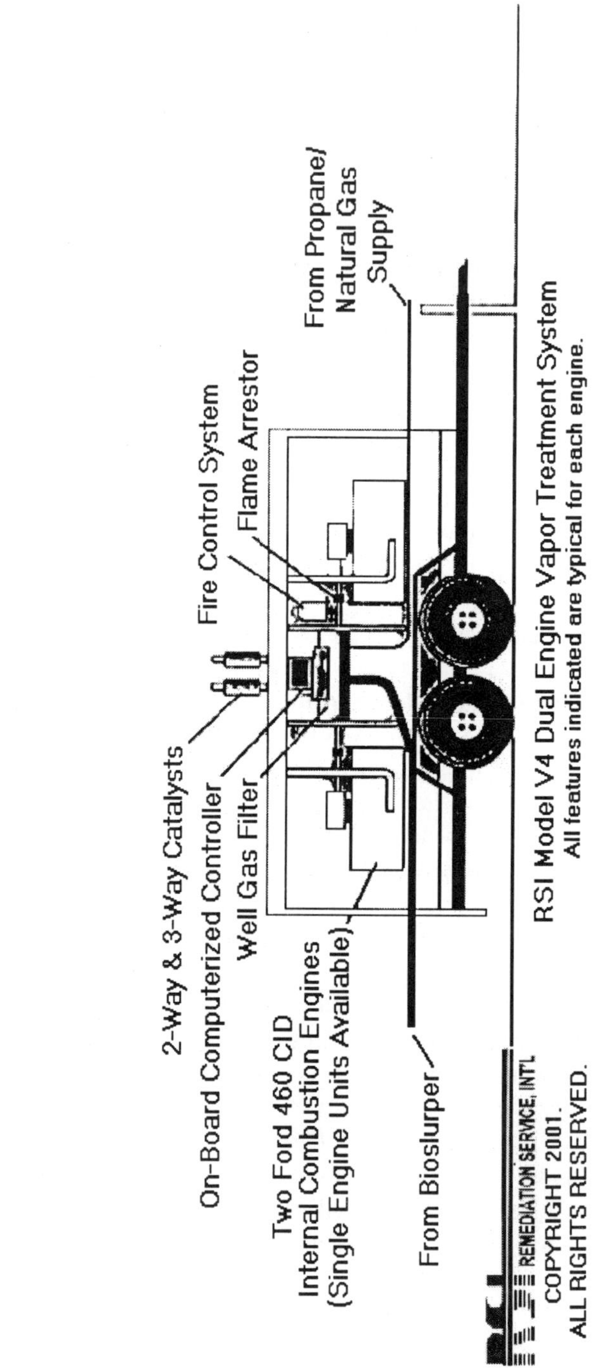

FIGURE C-1. Schematic diagram of the vapor treatment system (Reproduced courtesy of Remediation Service International, © 2001. All rights reserved.)

Catalysts have a finite life span (typically expressed in hours of operation) and must be monitored as that time approaches to ensure that they are working properly. The length of operation of the catalyst depends on the vapor concentration and whether lead or other potential catalyst poisons are present in the stack-gas contaminants. One equipment vendor suggested a range of 750 to 1,500 hours (about one to two months) of operation. A deactivated catalyst can be replaced easily with any automobile catalytic converter, available at most auto parts stores.

Data obtained from ICE operators and regulators show that ICEs are capable of destruction efficiencies of well over 99% (U.S. EPA, 1991). ICEs are especially useful for treating vapor streams with high concentrations of TPH (up to 30% volume) to levels below 50 ppmv. The vapor streams must be diluted with air to allow the ICE unit to treat stack gas containing more than about 16,000 ppmv of organics. ICEs also can effectively treat low concentrations (i.e., inlet vapor concentration below 1,000 ppm), although supplemental fuel use increases as the inlet concentration drops below 14,000 ppmv and the cost-effectiveness decreases at reduced intake concentrations. Tests of destruction of BTEX by ICE treatment show that nondetectable levels of contaminants can be achieved in the outlet stack gas in some cases, and outlet concentrations below 1 ppmv can be achieved in many cases.

ICEs offer advantages over conventional treatment methods, such as carbon adsorption, flame incineration, and catalytic incineration, at least for some applications. One advantage is the ability to produce power that can provide useful work output. Another advantage is their portability. Some disadvantages also have been noted. The primary drawback may be that the method requires a fairly high degree of manual supervision, especially during system startup. Mainly, the air-to-fuel ratio must be adjusted to maintain the proper conditions for complete combustion. Microcomputers are available to monitor and adjust the air-to-fuel ratio and add propane as needed; however, immediately after system startup, the characteristics of the extracted vapors may change so quickly that manual adjustment is required. The mechanical complexity of the ICE is another disadvantage. Over long-term operation, the maintenance costs for an ICE can be higher than for catalytic oxidation or flame incineration. Noise associated with the operation of the engine could be a concern in areas near residential zones or occupied buildings. Noise can be abated by adjusting the engine speed during certain time periods, installing a noise suppression fence, or purchasing special low-noise ICE models (AFCEE, 1994).

The capital cost of currently available ICE units appears to be somewhat higher, but certainly is in the same general range as for catalytic incineration and flame incineration. The capacities of four ICE treatment units are summarized in Table C-1. O&M costs are site-specific. In general, ICE costs appear to be somewhat higher than for other thermal processes, but generally they are in the same range as catalytic incineration and flame incineration. Because ICEs use a much more widely understood technology, gaining regulatory acceptance appears to be easier than for other technologies, and as a result, permitting and monitoring costs should be lower.

TABLE C-1. Capacities and costs of internal combustion engines for stack-gas treatment

Unit Size (Cubic-Inch Displacement)	Number of Cylinders	Maximum Capacity (scfm)
140	4	65
300	6	140
460	8	210
920	2 × 8	430

C.8 References

AFCEE, see Air Force Center for Environmental Excellence.

Air Force Center for Environmental Excellence. 1994. *A Performance and Cost Evaluation of Internal Combustion Engines for the Destruction of Hydrocarbon Vapors from Fuel-Contaminated Soils*. Prepared by S.R. Archabal and D.C. Downey. Environmental Services Office, Air Force Center for Environmental Excellence, Brooks Air Force Base, TX.

Buck, F.A.M., and C.W. Hauck. 1992. "Vapor Extraction and Catalytic Oxidation of Chlorinated VOCs." Proceedings of the 11th Annual Incineration Conference. Albuquerque, NM, May.

Connolley, M., B. Gibbs, and B. Keet. 1995. "Bioslurping Applied to a Gasoline and Diesel Spill in Fractured Rock." In R.E. Hinchee, J.A. Kittel, and H.J. Reisinger (Eds.), *Applied Bioremediation of Petroleum Hydrocarbons*, pp. 371-377. Battelle Press, Columbus, OH.

CSM Systems, 1989. Company literature. CSM Systems, Inc. Brooklyn, NY.

Downey, D.C., C.J. Pluhar, L.A. Dudus, P.G. Blystone, R.N. Miller, G.L. Lane, and S. Taffinder. 1994. "Remediation of Gasoline-Contaminated Soils Using Regenerative Resin Vapor Treatment and In Situ Bioventing." Proceedings of the Petroleum Hydrocarbons and Organic Chemicals Ground Water: Prevention, Detection, and Restoration Conference.

Kiang, Y.-H. 1988. "Catalytic Incineration." In Harry Freemen (Ed.), *Standard Handbook of Hazardous Waste Treatment and Disposal*. McGraw-Hill, New York, NY.

Little, A.D. 1987. *The Installation Restoration Program Toxicology Guide*, Vol. 3. Prepared for the Aerospace Medical Division. U.S. Air Force Systems Command, Wright-Patterson Air Force Base, Dayton, OH.

Remediation Service International. 2001. 4835 Colt Street, #D, Ventura, CA 93003 USA; rsi@rsi-save.com.

RSI, see Remediation Service International.

Rubin, D.K. 1995. "Industrial Cleanup Market Heats Up for Flameless Technology Vendor." *Engineering News Record, 235*(8):52-53.

Skladany, G.J., A.P. Togna, and Y. Yang. 1994. "Using Biofiltration to Treat VOCs and Odors." Superfund XIV Conference and Exhibition. Hazardous Materials Control Resources Institute, Rockville, MD.

U.S. Environmental Protection Agency. 1986. *Control Technologies for Hazardous Air Pollutants*. EPA/625/6-86/014.

U.S. Environmental Protection Agency. 1991. *Soil Vapor Extraction Technology: Reference Handbook.* EPA/540/2-91/003. Office of Research and Development, Risk Reduction Engineering Laboratory, Cincinnati, OH.

U.S. Environmental Protection Agency. 1994. *VISITT Vendor Information System for Innovative Treatment Technologies — User Manual (Version 3.0) with Update to Version 4.0.* EPA/542/R-94/003. Office of Solid Waste and Emergency Response, Washington, DC.

U.S. EPA, see U.S. Environmental Protection Agency.

Index

A

adsorption, 1, 145, 155, 156
 anthracite/clay, 88, 89, 111, 152
 carbon, 88, 89, 114, 145, 150, 163
 coefficient of soil, 20
 granular activated carbon (GAC), 89, 90, 92, 111, 155
aeration, 29, 72, 134
air
 distribution well, 157
 /LNAPL interface, 30, 43
 permeability, 20, 92
 pressure, 3, 16, 160
 saturation, 13
 /water interface, 10
anthracite/clay adsorption, 88, 89, 111, 152
aqueous phase, 10, 11, 57, 150
aquifer, 5, 12, 15, 32, 34, 35, 36, 68
 transmissivity, 15
atmospheric pressure, 59, 161

B

backpressure, 104, 105, 106, 111, 114
benzene, toluene, ethylbenzene, and xylenes (BTEX), 59, 60, 61, 62, 89, 130, 142, 163
biodegradation, 1, 18, 19, 24, 38, 62, 71, 72, 92, 113, 156, 157
bioslurper
 converting pilot-scale to full-scale, 137
 exit strategy, 119
  decision diagram for, 120
 feasibility evaluation, 23
bioslurper (*cont'd*)
 full-scale, 52, 66, 68, 75, 76, 85, 88, 128, 135, 136
 installation, 7
 liquid effluent, 114
 pilot-scale, 50, 51, 133, 156
 pilot test, 27, 42, 43, 68, 72
  permitting, 42
  shakedown checklist, 54
 shutdown, 72, 94, 101, 102, 107, 108, 115, 121, 161
 technology description, 6
 trailer, 51, 53, 124
bioslurping
 bioventing component of, 18
 cost-effectiveness, 41
 costs estimator, 123, 143
 installation, operations, and maintenance, 123
 methods for LNAPL recovery, 4
 SVE component of, 19
bioventing, 6, 18, 19, 38, 66, 71, 121
blower (pump), 20, 52, 78, 113, 114
boiling-point distribution, 37, 61, 62
BTEX, *see* benzene, toluene, ethylbenzene, and xylenes
bulk density, 60, 130

C

capillary
 forces, 7, 13
 fringe, 7, 15
 pressure, 12, 13, 15, 16, 20
 zone, 7, 58
capture zone, 66, 68
carbon adsorption, 88, 89, 114, 145, 150, 163

chemical reaction/flocculation (CRF), 109, 110, 145, 146, 148, 150
cone of depression, 5, 15, 55, 85
coprecipitation, 145
CRF, *see* chemical reaction/flocculation

D

DAF, *see* dissolved air flotation
density, 9, 13, 16, 17, 25, 37, 60, 68, 130
  LNAPL, 16, 25
  p-phase, 17
  water, 16, 17
desorption, 156
  soil, 20
dewatering, 14
  vacuum-enhanced, 18
differential pressure, 12, 58
discharge pressure (pump), 111, 152, 153
dissolved
  air flotation (DAF), 88, 89, 109, 110, 145, 148, 150
  hydrocarbons, 89, 114, 152
distribution
  boiling-point, 37, 61, 62
  LNAPL, 31, 68
  particle size, 60
  pore size, 13
  pressure, 68
  water, 88
drawdown, 1, 3, 5, 6, 7, 15, 16, 41, 55, 64, 65, 66, 71, 121
  dual-pump, 2, 53
  extraction, 15
  piezometric, 15
  pumping, 1, 5, 6, 7
  single-pump, 2
  test, 55, 64, 65
driving head, 5
drop tube, 39, 44, 55, 56, 57, 59, 64, 65, 66, 77, 80, 98, 100, 101
  dual, 84, 85
  optimum level, 101

E

emission(s)
  bioslurper, 127
  fume, 117
  limits, compound-specific, 90
  off-gas, 127, 136
  quantity (-ies), 90
  regulations, 90, 127
  source(s), 90
  surface testing, 92, 157
  testing, 113
emulsified hydrocarbons, 89
engine vacuum, 160, 161
exit strategy for bioslurping, 119
  decision diagram for, 120
extraction
  drawdown, 15
  groundwater, 14, 15, 108
  LNAPL, 1, 2, 15, 69, 95
  soil gas, 6, 18, 59, 80
  soil vapor (SVE), 2, 6, 19, 20, 21, 22, 55, 65, 66, 71, 72, 102, 165
  technologies, 1, 69, 95
  test, 55, 57, 69, 72, 80
  tube, 20, 139
  vacuum (vacuum-enhanced), 15, 18, 78
  well, 15, 16, 17, 33, 38, 39, 43, 44, 45, 46, 48, 52, 53, 57, 58, 62, 64, 65, 66, 71, 75, 77, 78, 80, 81, 82, 84, 85, 96, 104, 107, 116, 128, 134, 135

F

free product (*see also* light, nonaqueous-phase liquid [LNAPL]), 1, 6, 19, 25, 41, 44, 65, 66, 96, 101, 117, 118, 127, 131, 134
  removal, 1, 6
free-phase liquid, 1, 2
Freundlich isotherm, 11

G

GAC, *see* granular activated carbon
gas chromatography/flame ionization detection (GC/FID), 61
gasoline, 1, 10, 19, 27, 32, 59, 61, 88, 98, 110, 151, 159
GC/FID, *see* gas chromatography/flame ionization detection
granular activated carbon (GAC), 89, 90, 109, 111, 114, 151, 152, 155, 156
  adsorption, 90, 92, 111, 155
gravity head, 5
groundwater
  depression, 5, 45
  extraction, 14, 15, 108
  recovery rate, 6, 42, 45, 55, 59, 71, 83, 106, 112, 113

H

hydraulic
  conductivity, 5, 15, 34, 35, 100
    high, 5, 101
    low, 100
  gradient, 13, 14, 15, 34, 35
  head, 15, 35
hydrocarbon (*see also* total petroleum hydrocarbons [TPH]), 1, 2, 9, 10, 11, 12, 14, 16, 18, 20, 21, 27, 30, 31, 33, 40, 54, 62, 65, 73, 99, 111, 115, 117, 152, 153, 155, 156, 158, 159, 160, 161, 164
  degradation, 62
  dissolved, 89, 114, 152
  emulsified, 89
  petroleum, 1, 18, 23, 42, 57, 58, 62, 65, 82, 99, 116, 158
  removal, 14, 152, 153, 154
  sorption, 11
  vapors, 99, 116, 117, 160

I

ICE, *see* internal combustion engine
interface
  air/LNAPL, 30, 43
  air/water, 10
  in situ probe, 62, 63
  LNAPL/water, 12, 16, 65, 84
  oil/water, 24, 25, 29, 33, 34, 37, 53, 55, 69, 100, 119, 139
  probe, 54, 63, 129
  product/groundwater, 55
interfacial tension, 10, 12, 13, 86
internal combustion engine (ICE), 90, 93, 94, 98, 115, 116, 124, 126, 127, 151, 160, 161, 163, 164
  vacuum, 160, 161
intrinsic permeability, 20
in-well separation, 84, 85
isotherm, 11, 155

J

jet propellant (JP), 18, 19, 27, 32, 37, 110, 117, 151
JP, *see* jet propellant

K

knockout tank, 75, 82, 83

L

Langmuir isotherm, 11
laser-induced fluorescence (LIF), 31
LEL, *see* lower explosive limit
LIF, *see* laser-induced fluorescence
light, nonaqueous-phase liquid (LNAPL) (*see also* free product)
  /air interface, 30, 43
  /water interface, 12, 16, 65, 84
  density, 16, 25
  distribution, 31, 68
  extraction, 1, 2, 15, 69, 95
    device, 2
    pump, 15
    technologies, 1, 69, 95
    test, 69
  pressure, 16

light, nonaqueous-phase liquid (*cont'd*)
recovery, 2, 3, 5, 6, 7, 14, 15, 18, 20, 25, 38, 39, 41, 43, 44, 45, 50, 56, 65, 66, 71, 75, 78, 119, 121, 127, 132
bioslurping, pumping, and skimming methods for, 4
rate, 3, 5, 6, 25, 38, 39, 41, 43, 45, 65, 71, 119, 132
vacuum-assisted, 6
relative permeability of, 15
removal, 1, 25, 39, 41, 96
saturation, 20, 31
storage tank, 104, 112, 154
vapor-phase, 19
/water interface, 12, 16, 65, 84
zone, 67, 69
linear isotherm, 11
liquid
entrainment, 18, 20, 59
entrapment, 18
-extraction vacuum, 78
ring pump (LRP), 20, 50, 51, 52, 53, 54, 55, 59, 64, 65, 72, 75, 78, 80, 81, 82, 84, 85, 91, 94, 97, 98, 99, 101, 103, 104, 105, 106, 107, 112, 117, 125, 134, 136
operation of, 81, 98, 99, 106, 107
sizing of, 52, 59, 78, 80, 101, 134
troubleshooting guide, 104-105
saturation, 20
storage tank, 53
LNAPL, *see* light, nonaqueous-phase liquid
low vacuum, 107, 108
lower explosive limit (LEL), 92, 93, 158, 159
LRP, *see* liquid ring pump

M

MDL, *see* method detection limit
method detection limit (MDL), 59, 60, 61
methyl-*tert*-butyl ether (MTBE), 10
moisture content, 12, 25, 60, 62, 72, 149
MTBE, *see* methyl-*tert*-butyl ether

N

NAPL, *see* nonaqueous-phase liquid
National Pollutant Discharge Elimination System (NPDES), 42, 88
nonaqueous-phase liquid (NAPL), 1, 2, 13, 17, 68, 80
nonwetting fluid, 12, 13, 20
NPDES, *see* National Pollutant Discharge Elimination System

O

octanol/water partitioning coefficient, 11, 12, 20
off-gas, 39, 72, 75, 89, 124, 127, 130, 136, 138, 139, 155, 156, 157, 160, 161
O&G, *see* oil and grease
oil and grease (O&G), 109, 145, 148, 149
oil saturation, 31, 37
oil/water
interface, 24, 25, 29, 33, 34, 37, 53, 55, 69, 100, 119, 139
separator (OWS), 51, 52, 53, 54, 56, 59, 62, 65, 71, 75, 82, 86, 87, 88, 91, 97, 101, 106, 108, 109, 113, 117, 125, 145, 148, 150, 151, 154
organic
carbon partitioning coefficient, 11, 12
contaminants, 1, 118, 156, 157, 159, 160
vapor analyzer (OVA), 129, 130
OVA, *see* organic vapor analyzer
OWS, *see* oil/water separator

P

PAH, *see* polycyclic aromatic hydrocarbon
particle size distribution, 60
partitioning coefficient
  octanol/water, 11, 12, 20
  organic carbon, 11, 12
PEL, *see* Permissible Exposure Limit
permeability, 7, 12, 13, 32, 33, 35, 44, 48, 68
  air, 20, 92
  intrinsic, 20
  relative, 13, 14, 17, 68
    LNAPL, 15
    three-phase, 13
  soil, 13, 48, 90
  soil-gas, 44, 157
Permissible Exposure Limit (PEL), 117
permit (permitting), 42, 43, 52, 71, 88, 90, 97, 128, 138, 158, 164
personal protective equipment (PPE), 116, 117
petroleum hydrocarbons (*see* hydrocarbon, petroleum *and* total petroleum hydrocarbons [TPH])
piezometric drawdown, 15
polycyclic aromatic hydrocarbon (PAH), 31
polyvinyl chloride (PVC), 45, 58, 63, 84, 85, 117, 125, 126, 128, 134
pore
  size, 12, 13
  space, 7, 9, 12, 13, 35, 36
porosity, 20, 32, 33, 35, 36, 60, 68, 130
porous medium, 9, 12, 14, 35
PPE, *see* personal protective equipment
p-phase density, 17
precipitation, 34, 66, 71
  chemical, 88
  of calcium carbonate scale, 107
pressure
  air, 3, 16, 160
  atmospheric, 59, 161
  back, 104, 105, 106, 111, 114
  capillary, 12, 13, 15, 16, 20
  differential, 12, 58
  distribution, 68
  drop, 111
  equalization tank, 98, 107
  gauge, 30, 52, 111, 113, 114, 125
  gradient, 6, 10, 15, 68, 153
  head, 3, 17
  of the LNAPL, 16
  pump discharge, 111, 152, 153
  soil-gas, 14, 55, 58, 62, 69
  vapor, 10, 20, 37, 108
  water, 16
  zero-gauge, 16
process
  stream, 6, 152
  water, 42, 59, 71, 72, 75, 82, 83, 85, 86, 88, 97, 101, 102, 103, 110, 111, 112, 113, 145, 148, 150, 151, 152, 153, 154
    treatment, 75, 154
product/groundwater interface, 55
pump (*see also* skimmer), 1, 2, 3, 5, 7, 14, 15, 25, 27, 29, 30, 52, 53, 54, 55, 56, 64, 65, 66, 71, 75, 78, 80, 81, 91, 94, 98, 99, 100, 101, 103, 104, 105, 106, 107, 108, 110, 111, 112, 114, 125, 136, 152, 153, 154
  blower, 20, 52, 78, 113, 114
  diaphragm, 99, 110
  discharge pressure, 111, 152, 153
  drawdown, 2, 53, 64
  liquid ring (LRP), 20, 50, 51, 52, 53, 54, 55, 59, 64, 65, 72, 75, 78, 80, 81, 82, 84, 85, 91, 94, 97, 98, 99, 101, 103, 104, 105, 106, 107, 112, 117, 125, 134, 136
    operation of, 81, 98, 99, 106, 107

pump
liquid ring (*cont'd*)
sizing of, 52, 59, 78, 80, 101, 134
troubleshooting guide, 104-105
scale formation in, 72, 98, 104, 105, 107
sump, 99, 125, 145
vacuum, 6, 18, 59, 80, 81, 84, 129
water (aqueous effluent) transfer, 53, 54, 91, 94, 99, 106, 114
pumping
drawdown, 1, 5, 6, 7
methods for LNAPL recovery, 4
vacuum-enhanced, 14, 15
PVC, *see* polyvinyl chloride

R

radius of influence (ROI), 3, 38, 44, 58, 66, 67, 68, 69, 75, 77, 80, 127, 132, 134, 135, 157
bioslurper, 67, 69
test, 38
vadose zone, 57, 58, 59, 62, 67, 69, 72
RCRA, *see* Resource Conservation and Recovery Act
recovered (recovery of) LNAPL, 2, 3, 5, 6, 7, 14, 15, 18, 20, 25, 38, 39, 41, 43, 44, 45, 50, 56, 65, 66, 71, 75, 78, 119, 121, 127, 132
recovery
rate (groundwater), 6, 42, 45, 55, 59, 71, 83, 106, 112, 113
trench, 2, 3, 5
vacuum-enhanced, 7, 14, 15, 21, 68, 69
well, 2, 3, 5, 6, 15, 38, 43
regulations, 43, 59, 116, 119
and cost-effectiveness, 88
for emissions, 90, 127
relative permeability, 13, 14, 17, 68
LNAPL, 15
three-phase, 13
residual saturation, 12, 13, 68
Resource Conservation and Recovery Act (RCRA), 22, 155
ROI, *see* radius of influence

S

saturated zone, 7, 15, 121
saturation, 12, 13, 17, 31, 68
air, 13
liquid, 20
LNAPL, 20, 31
oil, 31, 37
residual, 12, 13, 68
water, 13, 20
scale formation in pumps, 72, 98, 104, 105, 107
skimmer (*see also* pump), 2, 3, 5, 25, 53, 55, 56, 64, 65, 112, 154
belt, 3, 110, 145, 148
drum, 3
floating filter scavenger system, 2
rope wick, 3
skimming, 1, 2, 6, 7, 41, 53, 66, 108, 109
methods for LNAPL recovery, 4
smear zone, 34
soil
adsorption, 20
desorption, 20
permeability, 13, 48, 90
vapor, 2, 65, 101, 103, 113
extraction (SVE), 2, 6, 19, 20, 21, 22, 55, 65, 66, 71, 72, 102, 165
soil gas, 6, 7, 9, 14, 16, 17, 18, 19, 20, 24, 27, 28, 29, 30, 38, 43, 44, 45, 46, 47, 48, 49, 52, 53, 55, 58, 59, 62, 65, 68, 69, 72, 78, 80, 81, 82, 85, 113, 116, 127, 128, 131, 132, 135, 138, 139, 142, 157
extraction, 6, 18, 59, 80
permeability, 44, 157
pressure, 14, 55, 58, 62, 69
vacuum, 68
/vapor, 6

sorption, 10, 11, 33, 37, 111, 156, 158
capacity, 156
coefficient, 11
curve, 111
of hydrocarbons, 11
isotherm, 11
source removal, 1, 66
specific gravity, 9, 17, 18
stack, 19, 39, 51, 52, 58, 59, 65, 66, 71, 75, 80, 90, 91, 92, 93, 96, 97, 98, 103, 113, 114, 115, 135, 156, 157, 158, 159, 163, 164
stack gas, 19, 39, 51, 52, 58, 59, 65, 66, 71, 75, 80, 90, 92, 93, 96, 97, 98, 103, 114, 115, 135, 156, 157, 158, 159, 163, 164
storage tank, 52, 53, 62, 82, 110, 114, 125, 126, 148, 156
fuel, 7, 52, 53, 62, 82, 91
liquid, 53
LNAPL, 104, 112, 154
superficial gas velocity (V_{SG}), 77, 78, 79
superficial liquid velocity (V_{SL}), 77, 79
surface
emissions, 92, 157
tension, 7, 12, 37, 68
surfactant, 12
SVE, *see* soil vapor extraction
Synthetic Precipitation Leaching Procedure, 119

T

TCLP, *see* Toxicity Characteristic Leaching Procedure
tension
interfacial, 10, 12, 13, 86
surface, 7, 12, 37, 68
three-phase
relative permeability, 13
system, 13, 15
zone, 9, 15, 17
threshold limit value (TLV), 117
TLV, *see* threshold limit value
total petroleum hydrocarbons (TPH) (*see also* hydrocarbon), 19, 27, 29, 38, 39, 42, 43, 45, 59, 60, 61, 62, 65, 71, 72, 80, 83, 85, 88, 90, 92, 93, 109, 113, 115, 116, 117, 119, 129, 130, 142, 148, 150, 156, 161, 163
total suspended solids (TSS), 61, 111, 148, 149, 152, 154
Toxicity Characteristic Leaching Procedure (TCLP), 110, 148
TPH, *see* total petroleum hydrocarbons
transmissivity, 15
trenches for vapor reinjection, 157
TSS, *see* total suspended solids
two-phase zone, 9, 17

U

underground storage tank (UST), 22, 119
unsaturated zone, 33, 68
UST, *see* underground storage tank

V

vacuum, 1, 6, 14, 15, 16, 17, 18, 19, 20, 33, 43, 44, 55, 57, 58, 59, 65, 68, 69, 78, 80, 81, 82, 84, 103, 104, 105, 107, 108, 113, 129, 136, 156, 160, 161
-assisted LNAPL recovery, 6
engine, 160, 161
-enhanced
bioslurper, 1
dewatering, 18
extraction, 15
pumping, 14, 15
recovery, 7, 14, 15, 21, 68, 69
extraction
liquid, 78
soil, 18
gauge, 125
gradient, 136
head, 69
loss, 107

vacuum (*cont'd*)
low, 107, 108
pump, 6, 18, 59, 80, 81, 84, 129
-resistant
separator, 83
tank, 82
soil-gas, 68
wellhead, 65, 80
vadose zone, 6, 7, 18, 24, 38, 39, 43, 48, 57-59, 62, 65, 66, 67, 69, 71, 72, 75, 92, 113, 121, 127
radius of influence (ROI), 57, 58, 59, 62, 67, 69, 72
remediation, 7, 66, 75
vapor, 6, 10, 14, 19, 20, 37, 42, 52, 53, 54, 55, 58, 59, 65, 66, 72, 80, 90, 91, 92, 93, 94, 97, 98, 99, 101, 107, 108, 110, 113, 114, 115, 116, 117, 130, 150, 151, 155, 156, 157, 158, 159, 160, 161, 162, 163, 164
analyzer, organic (OVA), 129, 130
concentration, 10, 90, 93, 115, 155, 156, 158, 159, 160, 163
discharge, 42, 52, 55, 58, 65, 66, 97
limits, 113
permits, 42, 97
pitot tube, 58
stack, 52
effluent, 113, 115
extraction from soil (SVE), 2, 6, 19, 20, 21, 22, 55, 65, 66, 71, 72, 102, 165
flow, 97, 108, 114
meter, 54
rate, 97
hydrocarbon, 99, 116, 117, 160
organic, 117, 118, 156, 159
phase, 19, 58, 65, 66, 110, 150
contaminants, 19
LNAPL, 19
modeling, 92
treatment, 155
vapor (*cont'd*)
pressure, 10, 20, 37, 108
recovery, 14
reinjection, 90
sample, 58, 59
sampling ports, 52, 113
soil, 2, 65, 101, 103, 113
soil gas, 6
stack, 115, 159
stream, 19, 92, 93, 114, 155, 156, 157, 158, 159, 163
TPH, 65
treatment, 42, 90, 91, 92, 93, 94, 97, 99, 113, 114, 151, 155, 162, 163
water, 114, 155, 156
wastestream, 97
viscosity, 10, 13, 37, 68
VOA, *see* volatile organic analysis
VOC, *see* volatile organic compound
volatile organic analysis (VOA), 61
volatile organic compound (VOC), 90, 102, 110, 150, 155, 158, 164, 165
volatility, 18, 27, 89
volatilization, 1, 10, 18, 19, 65
V_{SG}, *see* superficial gas velocity
V_{SL}, *see* superficial liquid velocity

W

wastewater treatment plant (WWTP), 42, 86, 88, 89, 103
water
density, 16, 17
distribution, 88
/LNAPL interface, 12, 16, 65, 84
/oil interface, 24, 25, 29, 33, 34, 37, 53, 55, 69, 100, 119, 139
pressure, 16
saturation, 13, 20
table, 1, 2, 5, 6, 7, 15, 19, 25, 27, 30, 33, 34, 35, 39, 43, 44, 48, 52, 55, 65, 66, 69, 71, 78

water (*cont'd*)
  transfer, 53, 54, 91, 94, 99, 106, 114
  treatment, 41, 51, 66, 75, 86, 88, 89, 91, 94, 97, 109, 124, 131, 154
  vapor, 114, 155, 156
well
  air distribution, 157
  casing, 34, 43, 44, 45, 55, 64
  diagram, 46
  diameter, 131
  extraction, 15, 16, 17, 33, 38, 39, 43, 44, 45, 46, 48, 52, 53, 57, 58, 62, 64, 65, 66, 71, 75, 77, 78, 80, 81, 82, 84, 85, 96, 104, 107, 116, 128, 134, 135
  layout, 77
  recovery, 2, 3, 5, 6, 15, 38, 43
  seal, 63
  screen (screened interval *or* section), 2, 43, 44, 45, 48, 50, 65, 71
wellhead vacuum, 65, 80
wettability, 12, 13
wetting fluid, 12, 13, 20
WWTP, see wastewater treatment plant

Z

zero-gauge pressure, 16
zone
  capillary, 7, 58
  capture, 66, 68
  LNAPL, 67, 69
  of influence, 68, 69, 132, 134
  saturated, 7, 15, 121
  smear, 34
  three-phase, 9, 15, 17
  two-phase, 9, 17
  unsaturated, 33, 68
  vadose, 6, 7, 18, 24, 38, 39, 43, 48, 57, 58, 59, 62, 65, 66, 67, 69, 71, 72, 75, 92, 113, 121, 127